Undergraduate Texts in Mathematics

Undergraduate Texts in Mathematics

Apostol: Introduction to Analytic Number Theory.
1976. xii, 338 pages. 24 illus.

Armstrong: Basic Topology.
1983. xii, 260 pages. 132 illus.

Bak/Newman: Complex Analysis.
1982. x, 224 pages. 69 illus.

Banchoff/Wermer: Linear Algebra Through Geometry.
1983. x, 257 pages. 81 illus.

Childs: A Concrete Introduction to Higher Algebra.
1979. xiv, 338 pages. 8 illus.

Chung: Elementary Probability Theory with Stochastic Processes.
1975. xvi, 325 pages. 36 illus.

Croom: Basic Concepts of Algebraic Topology.
1978. x, 177 pages. 46 illus.

Fischer: Intermediate Real Analysis.
1983. xiv, 770 pages. 100 illus.

Fleming: Functions of Several Variables. Second edition.
1977. xi, 411 pages. 96 illus.

Foulds: Optimization Techniques: An Introduction.
1981. xii, 502 pages. 72 illus.

Franklin: Methods of Mathematical Economics. Linear and Nonlinear Programming. Fixed-Point Theorems.
1980. x, 297 pages. 38 illus.

Halmos: Finite-Dimensional Vector Spaces. Second edition.
1974. viii, 200 pages.

Halmos: Naive Set Theory.
1974, vii, 104 pages.

Iooss/Joseph: Elementary Stability and Bifurcation Theory.
1980. xv, 286 pages. 47 illus.

Jänich: Topology
1984. ix, 180 pages (approx.). 180 illus.

Kemeny/Snell: Finite Markov Chains.
1976. ix, 224 pages. 11 illus.

Lang: Undergraduate Analysis
1983. xiii, 545 pages. 52 illus.

Lax/Burstein/Lax: Calculus with Applications and Computing, Volume 1.
1976. xi, 513 pages. 170 illus.

LeCuyer: College Mathematics with A Programming Language.
1978. xii, 420 pages. 144 illus.

Macki/Strauss: Introduction to Optimal Control Theory.
1981. xiii, 168 pages. 68 illus.

Malitz: Introduction to Mathematical Logic: Set Theory - Computable Functions - Model Theory.
1979. xii, 198 pages. 2 illus.

Martin: The Foundations of Geometry and the Non-Euclidean Plane.
1975. xvi, 509 pages. 263 illus.

Martin: Transformation Geometry: An Introduction to Symmetry.
1982. xii, 237 pages. 209 illus.

Millman/Parker: Geometry: A Metric Approach with Models.
1981. viii, 355 pages. 259 illus.

continued after Index

David R. Owen

A First Course in the Mathematical Foundations of Thermodynamics

With 52 Illustrations

Springer-Verlag
New York Berlin Heidelberg Tokyo

David R. Owen
Department of Mathematics
Carnegie-Mellon University
Pittsburgh, PA 15213
USA

AMS Subject Classification: 80-01

Library of Congress Cataloging in Publication Data
Owen, David R.
A first course in the mathematical foundations of thermodynamics.
(Undergraduate texts in mathematics)
Bibliography: p.
Includes index.
1. Thermodynamics. 2. Thermodynamics—Mathematics.
I. Title. II. Series.
QC311.094 1983 536'.7 83-14705

Softcover reprint of the hardcover 1st edition 1984

Typeset by Science Typographers, Medford, New York.

9 8 7 6 5 4 3 2 1
ISBN-13: 978-1-4613-9507-2 e-ISBN-13: 978-1-4613-9505-8
DOI: 10.1007/978-1-4613-9505-8

To Diane, Danny, Mom, and Dad

Preface

Research in the past thirty years on the foundations of thermodynamics has led not only to a better understanding of the early developments of the subject but also to formulations of the First and Second Laws that permit both a rigorous analysis of the consequences of these laws and a substantial broadening of the class of systems to which the laws can fruitfully be applied. Moreover, modern formulations of the laws of thermodynamics have now achieved logically parallel forms at a level accessible to undergraduate students in science and engineering who have completed the standard calculus sequence and who wish to understand the role which mathematics can play in scientific inquiry.

My goal in writing this book is to make some of the modern developments in thermodyamics available to readers with the background and orientation just mentioned and to present this material in the form of a text suitable for a one-semester junior-level course. Most of this presentation is taken from notes that I assembled while teaching such a course on two occasions. I found that, aside from a brief review of line integrals and exact differentials in two dimensions and a short discussion of *infima* and *suprema* of sets of real numbers, juniors (and even some mature sophomores) had sufficient mathematical background to handle the subject matter. Many of the students whom I taught had very limited experience with formal and rigorous mathematical exposition. For this reason, I have begun with a presentation of the foundations of classical thermodynamics in which many aspects of modern treatments are introduced in the concrete and simple setting of homogeneous fluid bodies. In addition, this material gives the student enough background to permit independent study of the vast classical literature and provides for an appreciation of the historical roots and logical structure of the modern treatments. I have benefited greatly from the beautiful exposition of classical thermodynamics by Truesdell and Bharatha[1], and my presentation of classical thermodynamics follows theirs in many respects. Because the main part of this book is itself an extension and broadening of the classical subject, I have chosen to give only an elementary, direct account of the classical foundations. In particular, I have treated a very limited subclass of the homogeneous fluid bodies covered by Truesdell and Bharatha.

Chapters II through V contain the basic material on the modern foundations of thermodynamics in the form of parallel treatments of the First and Second Laws. The laws of thermodynamics are first stated in terms of the concepts of heat, hotness, and work and mention only *special cycles* of a thermodynamical system. The main goal in these chapters is to express the content of these laws in terms of *arbitrary processes* of a system. This turns out to be more difficult in the case of the Second Law, because it is a non-trivial matter even to re-express its content in a form which applies to arbitrary cycles. The results of Chapters III and IV establish the equivalence of the First and Second Laws with statements of the following type: *a distinguished interaction between a system and its environment either vanishes or is of fixed sign in every cycle*. For the First Law, this statement expresses the fact that thermal and mechanical interactions are comparable in a simple and precise sense. For the Second Law, this statement is a generalization of the Clausius inequality given in traditional treatments of thermodynamics and, in rough terms, says that the integral of the heat added divided by absolute temperature is never positive for a cycle. It is important to note here that the notion of an absolute temperature scale emerges from the analysis leading to this equivalent statement of the Second Law and permits one to free the original statement of the Second Law from its restriction to special cycles. A similar remark applies to the notion of "the mechanical equivalent of heat" with respect to the First Law.

The main goal of the modern development is fully realized in Chapter V, where statements of the type in italics above are shown to be equivalent to the existence of functions of state whose differences are equal to or provide upper bounds for the distinguished interactions. In the case of the First Law, equality holds and the function of state is called an energy function for the system; for the Second Law, the function of state, or "upper potential", is called an entropy function. At bottom, the modern work on foundations shows that the concepts of absolute temperature, energy, and entropy can be obtained as consequences of primitive, natural, and general statements of the First and Second Laws, and such concepts yield equivalent statements of these laws which are known to be important for applications.

[1] Classical Thermodynamics as a Theory of Heat Engines, 1977, New York Heidelberg Berlin: Springer-Verlag.

As I mentioned above, the study of homogeneous fluid bodies in Chapter I helps to ease the reader into the more abstract and formal modern work, and homogeneous fluid bodies appear extensively in Chapters II through V both as illustrations and as important "model systems". Because modern applications include many systems not describable as homogeneous fluid bodies, I have devoted the bulk of Chapters VI and VII to a discussion of the main features of some important non-classical systems: viscous filaments, elastic–perfectly plastic filaments, and homogeneous bodies with viscosity. In order to keep the discussion at the right mathematical level and to permit easy comparison with the earlier analysis of homogeneous fluid bodies, I chose to maintain the classical assumption of spatial homogeneity. The additional restriction to isothermal processes made in Chapter VI provides further simplification: it yields a statement of the Second Law which mentions only the concept of work. In Chapter VI, elastic filaments are the counterparts of homogeneous fluid bodies, in that every process of an elastic filament has a reversal, and the work action changes signs under reversals. Viscous filaments do not have these properties, but instead exhibit approximate behavior of this type when processes are performed sufficiently slowly. Elastic–perfectly plastic filaments are similar to viscous filaments, in that they do not admit reversals of arbitrary processes. However, they differ from viscous filaments in an important way: there is no approximation for elastic–perfectly plastic behavior which yields a classical analysis of reversals.

I regard the study of systems with viscosity as crucial for a thorough understanding of thermodynamics, because these systems provide a simple and precise way of making explicit one of the kinds of approximations which underlie the concept of "quasi-static processes" in traditional presentations. Therefore, I have devoted a second section of Chapter VI as well as all of Chapter VII to systems with viscosity. In Section 4 of Chapter VI, I begin with the notion of the latent heat associated with a phase transition at fixed temperature in an elastic filament. Although the latent heat is not well defined for an analogous process in a viscous filament, it is natural to consider an elastic filament which approximates the viscous filament in slow processes and to compare its latent heat with the heat gained by the viscous filament. The results of this analysis give a vivid illustration of the lack of symmetry in the roles which thermal and mechanical energy play in nature: work done against viscous forces in a filament can contribute to melting or vaporization, but solidification and condensation cannot cause viscous forces in the filament to do work on its surroundings. In Chapter VII, I remove the restriction that processes be isothermal and study the consequences of the First and Second Laws for homogeneous viscous bodies. The modern conceptual framework is employed to analyze viscous bodies, and this analysis features homogeneous fluid bodies as approximating systems. The material in Chapter VII helps the reader to view the results of Chapter I on classical thermodynamics from a modern perspective and establishes the consistency of the classical and modern approaches.

The reader will notice that this book is not self-contained in two respects. First of all, I have not attempted to provide here the background material from elementary mathematical analysis used throughout the book. This material includes standard notions of continuity, differentiability, and integrability (in Riemann's sense) for real-valued functions of no more than three variables. Secondly, I have not discussed the background for the concepts of heat and hotness in a way which emphasizes the physical underpinnings of these notions. Such a discussion would fall under the headings of calorimetry and thermometry in standard physics texts on thermodynamics. For both types of background material, I have provided references at the end of this book. Also, I have there outlined the contributions of various researchers to modern thermodynamics and have provided references to some of their work. In the outline and references, no attempt has been made to survey all of the interesting and fruitful areas of modern research in thermodynamics. Indeed, attention is focused there on work closely related to the approach taken in this book. Nevertheless, some of the references given there will help the reader who is interested in other directions of modern research.

It is a pleasure to acknowledge the inspiration and encouragement provided through the years by my teachers and colleagues Bernard D. Coleman, Morton Gurtin, Walter Noll, James Serrin, and Clifford Truesdell. It is my hope that this book adequately reflects not only their specific contributions to thermodynamics but also a common goal underlying their scientific research: the realization of a harmonious union of natural science and mathematics. I wish also to thank John Thomas for his helpful comments at various stages in the preparation of this book and Stella DeVito for her expert typing of the manuscript. Finally, I am indebted to Ernest MacMillan and K. R. Rajagopal for their careful reading of the galleys and page proofs.

Contents

Introduction

Thermodynamics studies the mechanisms through which physical systems become hotter or colder. Such mechanisms include conduction of electricity, mechanical friction, thermal radiation, and surface heat transfer. One of the important tasks of thermodynamics is to explore the restrictions which nature places on these mechanisms. An example of a restriction of this type is obtained by considering a metal block at rest on a rough metal table with both block and table at the same temperature T_1. Our physical experience tells us that the two objects will not in and of themselves supply thermal energy to their common interface with the effect that the block accelerates from rest relative to the table, and the block and table both attain a lower temperature T_0. On the other hand, our experience does accept as plausible the reverse sequence of events: both objects begin at temperature T_0 with the block moving relative to the table and then the block decelerates to rest while the objects heat up to temperature T_1. This example suggests that relative motion of rough surfaces in contact can cause objects to become hotter, but that objects do not spontaneously cool down in order to initiate such relative motion. In other terms, the initial mechanical energy of the moving block can cause thermal energy to flow into the block and the table, whereas that same thermal energy of the rest system is not directly available for supplying mechanical energy to the block. Thus, mechanical and thermal energy appear not to play symmetrical roles in nature, and other elementary examples show this to be the case of electrical and thermal energy. The main goal of thermodynamics is to make such restrictions explicit and to analyze the consequences of these restrictions.

Much of classical thermodynamics concerns thermal and mechanical interactions between a physical system and its environment. In fact, when thermodynamics emerged in the last century as a science, there was much effort devoted to the problem of designing efficient engines which could deliver work through the heating and cooling of an operating medium such as water, and the celebrated Second Law of thermodynamics itself arose out of attempts to maximize the efficiency of heat engines. Throughout this presentation, thermal and mechanical interactions will be studied to the exclusion of electromagnetic and chemical interactions. Thus, we may call the subject of this book *thermomechanics* and content ourselves with the fact that a solid understanding of this subject will provide a basis for the study of other branches of thermodynamics.

The physical concepts underlying thermomechanics turn out to be those of heat, hotness and work, and these concepts are at the heart of both classical and modern treatments. Where classical and modern approaches to thermomechanics differ is in the nature of the physical systems which they attempt to study. One of the major aims of this book is that of making these differences explicit. For the moment it suffices to say only that modern treatments of thermomechanics cover a much broader class of physical systems than do the classical treatments.

The restrictions on interacting systems which form the basis of thermodynamics usually begin as observations about simple physical systems and are subsequently restated as general laws which are required to hold "universally." Much of the difficulty in traditional treatments of thermodynamics lies in a failure to make explicit the meaning of the word "universally." To give this word a clear meaning and thereby to delimit the scope of thermodynamics requires the use of a precise language to describe basic concepts, to state assumptions, and to carry out logical arguments. Mathematics is the most appropriate language for this purpose, and the reader will note from the outset that this presentation uses mathematics more extensively than most. Indeed, this presentation is axiomatic in nature, as are the standard presentations of Euclidean geometry. In order to prevent discussions from becoming unduly formal, I shall try to provide whenever possible physical motivation for the assumptions made and physical interpretations for the results obtained. Thus, my goal is to present some basic aspects of thermodynamics by allowing physical ideas to provide the basic input and mathematical structure to provide the vehicle for expressing and developing these ideas.

List of Symbols

$\mathscr{F}$	homogeneous fluid body	I-1
Σ	state space	I-1
V	volume	I-1
θ	temperature	I-1
$\not{p}$	pressure function	I-1
$\tilde{\lambda}$	latent heat function	I-1
$\mathscr{c}$	specific heat function	I-1
$\mathbb{R}$	real numbers	I-1
$\mathbb{R}^{++}$	$(0, \infty)$ = positive real numbers	I-1
$\mathbb{P}$	path	I-1
$(\bar{V}, \bar{\theta})$	standard parameterization	I-1
$W(\mathbb{P})$	work done along $\mathbb{P}$	I-1
$H(\mathbb{P})$	net heat gained along $\mathbb{P}$	I-1
$\bar{\mathrm{h}}(\tau)$	heating at time τ	I-1
$H^{+}(\mathbb{P})$	heat absorbed along $\mathbb{P}$	I-1
$H^{-}(\mathbb{P})$	heat emitted along $\mathbb{P}$	I-1
$\mathbb{P}_r$	reversal of $\mathbb{P}$	I-1
$\mathbb{P}_2 * \mathbb{P}_1$	$\mathbb{P}_1$ followed by $\mathbb{P}_2$	I-1
$\mathrm{Int}\,\mathbb{P}$	interior of $\mathbb{P}$	I-1
$\mathrm{J}, \mathrm{R}, \lambda$	constants for an ideal gas	I-1
$\mathscr{G}$	ideal gas	I-1
E	energy function	I-2
J	mechanical equivalent of heat	I-2
$\check{W}$	work done on $\mathscr{F}$	I-2
$\mathbb{C}$	Carnot heat engine	I-3
S	entropy function	I-3

E°	normalized energy function	I-3
S°	normalized entropy function	I-3
Π	set of process generators	II-1
π	process generator	II-1
ρ_π	transformation induced by π	II-1
$\mathscr{D}(\pi)$	domain of ρ_π	II-1
$\mathscr{R}(\pi)$	range of ρ_π	II-1
$\Pi\sigma$	set of states accessible from σ	II-1
$\pi''\pi'$	π' followed by π''	II-1
$\emptyset$	empty set	II-1
$\rho_{\pi'}^{-1}$	inverse image function for $\rho_{\pi'}$	II-1
$\Pi \Diamond \Sigma$	set of processes of (Σ, Π)	II-1
$(\Sigma_{\mathscr{F}}, \Pi_{\mathscr{F}})$	system with perfect accessibility corresponding to a homogeneous fluid body $\mathscr{F}$	II-1
π_t	a process generator for $(\Sigma_{\mathscr{F}}, \Pi_{\mathscr{F}})$	II-1
$\Gamma(\pi_t, \sigma)$	path determined by (π_t, σ)	II-1
$(\Pi \Diamond \Sigma)_{\text{cyc}}$	the set of cycles of (Σ, Π)	II-1
a	an action	II-2
a_H	the action determined by the heat gained H	II-2
$\mathscr{S}$	thermodynamical system	III-1
$(\Sigma_{\mathscr{S}}, \Pi_{\mathscr{S}})$	system with perfect accessibility corresponding to a thermodynamical system $\mathscr{S}$	III-1
$W_{\mathscr{S}}$	work action for $\mathscr{S}$	III-1
$H_{\mathscr{S}}$	heat action for $\mathscr{S}$	III-1
$(\Pi_{\mathscr{S}} \Diamond \Sigma_{\mathscr{S}})_{\text{cyc}}$	the set of cycles of $\mathscr{S}$	III-1
$\mathscr{S}_1 \times \mathscr{S}_2$	product of $\mathscr{S}_1$ and $\mathscr{S}_2$	III-2
$\mathbb{U}$	a collection of systems	III-3
$\mathscr{M}$	hotness manifold	IV-2
$\prec$	order on hotness manifold	IV-2
L	hotness level	IV-2
$\mathscr{T}$	set of empirical temperature scales	IV-2
φ	empirical temperature scale	IV-2
$\varphi_{\mathscr{G}}$	empirical temperature scale for an ideal gas $\mathscr{G}$	IV-3
$H_{\mathscr{G}}$	accumulation function for an ideal gas $\mathscr{G}$ (as a function of hotness)	IV-3
$\hat{H}_{\mathscr{G}}$	accumulation function for an ideal gas $\mathscr{G}$ (as a function of temperature)	IV-3
$\mathbf{A}$	the adiabat in the proof of the Lemma	IV-3
$H_{\mathscr{S}}$	accumulation function for a thermodynamical system $\mathscr{S}$	IV-4
L′, L″	lower and upper hotness levels for a process	IV-4
$H_{\mathscr{S}}(\pi, \sigma, \mathrm{L})$	net heat gained by $\mathscr{S}$ in the process (π, σ) at or below L	IV-4
$H_{\mathscr{S}}(\pi, \sigma)$	net heat gained by $\mathscr{S}$ in the process (π, σ)	IV-4

$H_{\mathscr{S}}^{\downarrow}$	net heat absorbed	IV-7
$H_{\mathscr{S}}^{\uparrow}$	net heat emitted	IV-7
A	upper potential for an action	V-2
E	energy function	V-3
$\check{W}_{\mathscr{S}}$	work done on $\mathscr{S}$ by its environment	V-4
Ψ	Helmholtz free energy function	V-4
ℓ	length of a homogeneous filament	VI-1
f	external force on a filament	VI-1
(ℓ_a, ℓ_b)	interval of admissible lengths of a filament	VI-2
ϕ_{e}	response function of an elastic filament	VI-2
$\mathscr{w}_{\mathrm{e}}$	work action for an elastic filament	VI-2
k	constant for a linear spring	VI-2
r	rate of change of length of a viscous filament	VI-3
ϕ_{v}	response function of a viscous filament	VI-3
$\mathscr{w}_{\mathrm{v}}$	work action for a viscous filament	VI-3
$\mu(\ell)$	viscosity of a linear viscous filament	VI-3
$\tilde{\ell}_{\mathrm{n}}$	retardation of a process of a filament	VI-3
ϕ_{eq}	equilibrium response function for a viscous filament	VI-3
Φ_{eq}	antiderivative of ϕ_{eq}	VI-3
$\Phi_{\mathrm{v}}^{\circ}$	normalized Helmholtz free energy function of a viscous filament	VI-3
f_y	yield force of an elastic-plastic filament	VI-4
β	elastic modulus of an elastic-plastic filament	VI-4
ℓ_{R}	residual length	VI-4
$\mathscr{w}_{\mathrm{ep}}$	work action for an elastic-plastic filament	VI-4
$\mathscr{L}^{\circ}$	the class of normalized Helmholtz free energy functions for an elastic-plastic filament	VI-4
$\hat{\Psi}^{\circ}, \Psi^{\circ}$	smallest and largest Helmholtz free energy functions	VI-4
$\mathscr{h}_{\mathrm{e}}$	heat action for an elastic filament	VI-5
f_{tr}	force at which a phase transition occurs	VI-5
λ_{12}	latent heat	VI-5
$\mathscr{h}_{\mathrm{v}}$	heat action for a viscous filament	VI-5
E_{v}	internal energy function of a viscous filament	VI-5
$\mathscr{V}$	homogeneous body with viscosity	VII-1
$\tilde{\mu}$	response function for a viscous body	VII-1
Γ_{eq}	equilibrium path	VII-1
$(\Sigma_{\mathscr{V}}, \Pi_{\mathscr{V}})$	system with perfect accessibility for a viscous body $\mathscr{V}$	VII-1
$\mathrm{W}_{\mathscr{V}}$	work action for a viscous body	VII-1
$H_{\mathscr{V}}$	accumulation function for a viscous body	VII-1
$\mathrm{I}_{\mathscr{V}}$	accumulation integral for a viscous body	VII-1

CHAPTER I
Classical Thermodynamics

1. Homogeneous Fluid Bodies

The physical systems which we shall study in this chapter are fluid bodies such as a definite quantity of a liquid or gas in a finite container. The term "homogeneous" is used to indicate that the state of the fluid does not vary from point to point within the container. It is not difficult to accept this notion of homogeneity as being both plausible and useful. What is difficult to accept, and what causes much confusion in both the exposition and application of classical thermodynamics, is the idea that a fluid body evolve in time and also be homogeneous at every instant. An evolution of this type is difficult to visualize, because we usually think of effecting changes in a body by manipulating its boundary in some way. (Here we can think of supplying mechanical or thermal energy to part of the container by moving one of its walls or placing hotter or colder bodies in contact with it.) We expect that the state of the fluid inside the container will change first at points near the container walls and later at points in the center of the container, and at any instant the condition of homogeneity will not strictly apply. Therefore, the notions of evolution in time and spatial homogeneity generally are not physically compatible. Nevertheless, the idea that a fluid body evolve through spatially homogeneous states is an essential part of classical thermodynamics, and we are led to interpret evolution of this type as an *approximation* to evolution which proceeds slowly compared to the speed at which local changes of state can propagate from point to point.

Thus, in classical thermodynamics it is natural to regard "evolution through homogeneous states" as an approximation to, or as an idealized description of, evolution which is slow (in a sense that depends upon the particular fluid body under consideration). It is interesting to look ahead momentarily and to note that, in our mathematical description of classical thermodynamics, the principal interactions between a homogeneous fluid body and its environment are independent of the speed at which the state of the fluid changes. In other words, the formal structure of classical thermodynamics turns out to be "rate-independent" and, therefore, cannot by itself offer an analysis of the approximations employed in its physical interpretation. A principal motivation for modern research in thermodynamics is that of providing a mathematical framework in which an analysis of this type can be carried out. Although the details of this analysis require more advanced mathematics than is used in this book, the mathematical foundations studied in Chapters II through V do provide a conceptual basis for such an analysis. Moreover, the examples of viscous behavior studied in Chapters VI and VII permit an explicit analysis of the effects of "retarding" processes in time and so allow us to describe precisely how a homogeneous fluid body can approximate a homogeneous body with viscosity.

We turn now to the basic concept in classical thermodynamics.

Definition 1.1. A *homogeneous fluid body* $\mathscr{F}$ is prescribed by giving a *state space* Σ, consisting of *states* (V, θ), a *pressure function* p, a *latent heat function* $\tilde{\lambda}$, and a *specific heat function* σ which satisfy the following conditions:

($\mathscr{F}$1). Σ is an open, convex subset of $\mathbb{R}^{++} \times \mathbb{R}^{++}$;
($\mathscr{F}$2). p, $\tilde{\lambda}$ and σ are real-valued and continuously differentiable on Σ;
($\mathscr{F}$3). throughout Σ, $\partial p/\partial V$ is negative, σ is positive, and $\tilde{\lambda}$ does not vanish.

We call V the *volume* and θ the *temperature* of the fluid in the state (V, θ). Since both V and θ are required in ($\mathscr{F}$1) to be positive numbers, we can visualize the state space Σ as a subset of the first quadrant of a V-θ plane, with Σ not touching either of the coordinate axes. We interpret θ as the temperature measured with respect to a preassigned scale. The choice of scale is fixed throughout this chapter and is restricted in a way which will be explained in Chapter IV. For definiteness, the reader may take the preassigned scale to be the Kelvin scale, which is obtained from the familiar Celsius scale by a shift of about 273 degrees, but neither an understanding of this particular scale nor an understanding of the concept of temperature scales, in general, is required in this chapter. The numbers $p(V, \theta)$, $\tilde{\lambda}(V, \theta)$, and $\sigma(V, \theta)$ are called the *pressure*, the *latent heat* (*with respect to volume*), and the *specific heat* (*at constant volume*), respectively, *at the state* (V, θ). [The symbols k and c_V often replace σ and the term "heat capacity" often replaces "specific heat" in traditional treatments.] As we shall indicate presently, the functions p, $\tilde{\lambda}$ and σ permit us to define the work and the net heat gained associated with certain physical processes of a homogeneous fluid body. In fact, these three functions describe all of the physical properties of a homogeneous fluid body which are relevant to classical thermodynamics.

The condition that $\partial p / \partial V$ always be negative means that, at constant temperature, the pressure in the fluid must increase as the volume decreases. The opposite condition $\partial p / \partial V > 0$ would mean that the force required to move a piston into a chamber filled with the fluid would decrease as the piston moves into the chamber and compresses the fluid. Thus, by simply putting a weight on top of the piston, we could compress the fluid down to zero volume (provided we can keep the temperature constant). This unstable situation is ruled out by ($\mathcal{F}$3). The condition that $\tilde{\lambda}$ not vanish and the continuity of $\tilde{\lambda}$ imply that $\tilde{\lambda}$ is of one sign on Σ. In later sections we will be in a position to show that this requirement on $\tilde{\lambda}$ excludes from our theory the so-called "anomalous" behavior of water at 4°C temperature, 1 atm pressure. In other words, our theory is not equipped to study water under these conditions.[1] The condition that σ be positive will be interpreted following our definitions of heat gained and work.

Classical thermodynamics is concerned with the evolution of physical systems in time, and this aspect of our treatment of homogeneous fluid bodies is described via the concept of a "path." A path represents the states which a body assumes during a time interval, taking into account the order in time in which the states are assumed but not the speed at which they are encountered. The next definition makes this concept precise.

Definition 1.2. A *path* $\mathbb{P}$ for a homogeneous fluid body $\mathcal{F}$ is an oriented, piecewise continuously differentiable curve in Σ.

The standard mathematical definition of the term "oriented curve" permits us to describe a path as a collection of "equivalent" functions mapping intervals of the form $[t_1, t_2]$ into Σ. Each such function can be thought of as determining a point $(V(\tau), \theta(\tau))$ which moves through Σ as τ increases from t_1 to t_2. Two functions of this type are then said to be equivalent if the moving points in each case trace out the same locus in Σ in the same direction, irrespective of the rate at which that locus is traversed. It turns out that we can specify a path $\mathbb{P}$ by giving a single function $(\bar{V}, \bar{\theta})$: $[0,1] \to \Sigma$ (Figure 1). A function of this form whose components have derivatives which, except at a finite number of times, do not vanish simultaneously is called a *standard parameterization* of $\mathbb{P}$.

In defining the concept of a path, we have ignored the speed at which a locus in Σ is traversed. This approach can be justified by the fact that, for homogeneous fluid bodies, the classical expressions for the work and the heat gained do not depend upon that speed.

[1]Truesdell and Bharatha allow $\tilde{\lambda}$ to vanish at certain points and require only that Σ be open and connected.

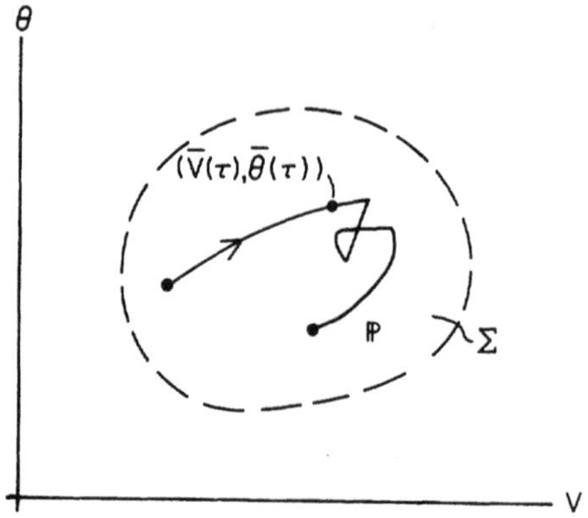

Figure 1. A path for $\mathscr{F}$.

Definition 1.3. Let $\mathbb{P}$ be a path for a homogeneous fluid body $\mathscr{F}$. The *work done by $\mathscr{F}$ along* $\mathbb{P}$ is the real number

$$W(\mathbb{P}) := \int_{\mathbb{P}} p\, dV$$

$$:= \int_0^1 p(\bar{V}(\tau), \bar{\theta}(\tau)) \bar{V}^{\cdot}(\tau)\, d\tau; \tag{1.1}$$

the *net heat gained by $\mathscr{F}$ along* $\mathbb{P}$ is the real number

$$H(\mathbb{P}) := \int_{\mathbb{P}} \tilde{\lambda}\, dV + \kappa\, d\theta$$

$$:= \int_0^1 [\tilde{\lambda}(\bar{V}(\tau), \bar{\theta}(\tau)) \bar{V}^{\cdot}(\tau) + \kappa(\bar{V}(\tau), \bar{\theta}(\tau)) \bar{\theta}^{\cdot}(\tau)]\, d\tau. \tag{1.2}$$

Here, $(\bar{V}, \bar{\theta})$ is a standard parameterization for $\mathbb{P}$, and $(\bar{V}^{\cdot}, \bar{\theta}^{\cdot})$ denotes its derivative.

The expressions for $W(\mathbb{P})$ and $H(\mathbb{P})$ in (1.1) and (1.2) as integrals over $[0,1]$ justify our use of the symbols $\int_{\mathbb{P}} p\, dV$ and $\int_{\mathbb{P}} \tilde{\lambda}\, dV + \kappa\, d\theta$, since the integrals over $[0,1]$ are, in fact, line integrals of the vector fields $(p, 0)$ and $(\tilde{\lambda}, \kappa)$ along the oriented curve $\mathbb{P}$. Our assertion above, that the classical expressions for the work and heat gain do not depend upon the speed of traversal of a locus in Σ, can now be justified by citing a standard result on the independence of parameterization for line integrals of vector fields. In fact, this result must also be used to show that the expressions for $W(\mathbb{P})$ and $H(\mathbb{P})$ do not depend upon the choice of standard parameterization $(\bar{V}, \bar{\theta})$ for $\mathbb{P}$ which appears in (1.1) and (1.2).

The formula (1.2) permits us to interpret the condition on κ in ($\mathscr{F}$3), Definition 1.1. According to (1.2), along any path with $\bar{V}$ a constant function, i.e. along any vertical path in Σ, $H(\mathbb{P})$ is given by the integral

$\int_{\mathbb{P}} \sigma \, d\theta$ or, equivalently,

$$H(\mathbb{P}) = \int_0^1 \sigma(\bar{V}(\tau), \bar{\theta}(\tau)) \bar{\theta}^{\cdot}(\tau) \, d\tau. \tag{1.3}$$

Therefore, since σ is assumed to be positive, $H(\mathbb{P})$ will be positive if θ increases along $\mathbb{P}$ and negative if θ decreases along $\mathbb{P}$. In other words, (1.2) and ($\mathscr{F}$3) require that a homogeneous fluid body *absorb* heat when its temperature is increasing at constant volume and *emit* heat when its temperature is decreasing at constant volume (Figure 2).

Let $(\bar{V}, \bar{\theta})$ be a standard parameterization of a path $\mathbb{P}$ for $\mathscr{F}$. The number

$$\bar{\mathrm{h}}(\tau) := \tilde{\lambda}(\bar{V}(\tau), \bar{\theta}(\tau)) \bar{V}^{\cdot}(\tau) + \sigma(\bar{V}(\tau), \bar{\theta}(\tau)) \bar{\theta}^{\cdot}(\tau) \tag{1.4}$$

is called the *rate of gain of heat* or the *heating at time* τ with respect to $(\bar{V}, \bar{\theta})$. Unlike the net heat gained along $\mathbb{P}$, the heating depends upon $(\bar{V}, \bar{\theta})$ as well as upon $\mathbb{P}$. The following definition gives integrals of certain functions of $\bar{\mathrm{h}}$ which, like $H(\mathbb{P})$, turn out to depend upon $\mathbb{P}$ alone.

Definition 1.4. The *heat absorbed* $H^+(\mathbb{P})$ *by* $\mathscr{F}$ *along* $\mathbb{P}$ and the *heat emitted* $H^-(\mathbb{P})$ *by* $\mathscr{F}$ *along* $\mathbb{P}$ are defined by

$$H^+(\mathbb{P}) := \frac{1}{2} \int_0^1 (\bar{\mathrm{h}}(\tau) + |\bar{\mathrm{h}}(\tau)|) \, d\tau \tag{1.5}$$

and

$$H^-(\mathbb{P}) := \frac{1}{2} \int_0^1 (-\bar{\mathrm{h}}(\tau) + |\bar{\mathrm{h}}(\tau)|) \, d\tau. \tag{1.6}$$

The integrands in (1.5) and (1.6) are non-negative functions of τ. In fact, $(\bar{\mathrm{h}}(\tau) + |\bar{\mathrm{h}}(\tau)|)/2$ equals $\bar{\mathrm{h}}(\tau)$ or zero depending upon whether $\bar{\mathrm{h}}(\tau)$ is positive or non-positive. Similarly, $(-\bar{\mathrm{h}}(\tau) + |\bar{\mathrm{h}}(\tau)|)/2$ equals zero or $-\bar{\mathrm{h}}(\tau)$, depending upon whether $\bar{\mathrm{h}}(\tau)$ is positive or non-positive (Figure 3).

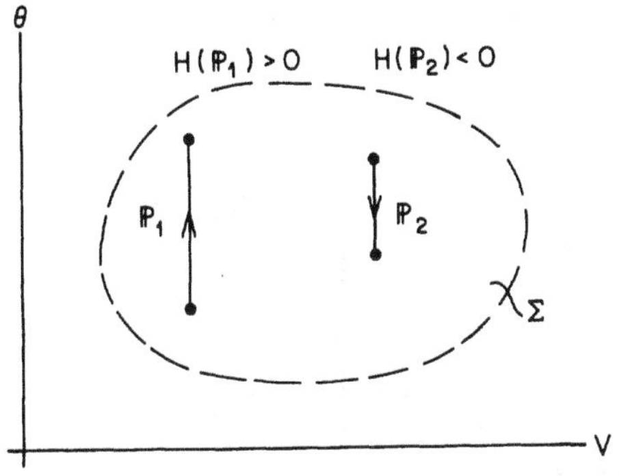

Figure 2. Paths of constant volume.

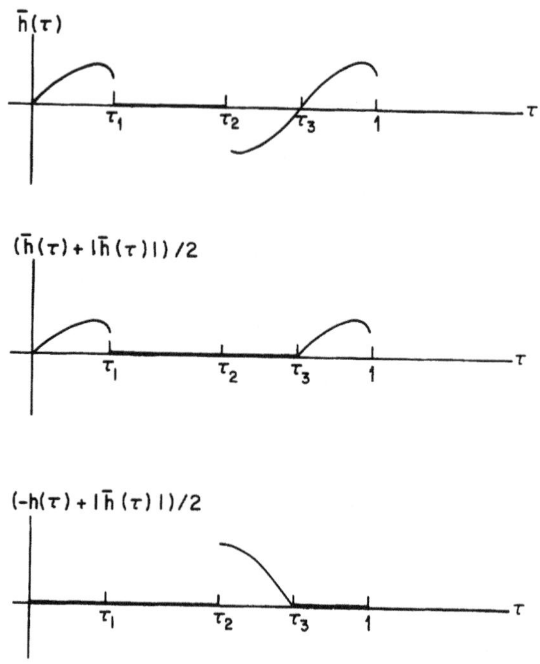

Figure 3.

The relations (1.2), (1.4), (1.5), and (1.6) immediately imply

$$H(\mathbb{P}) = H^{+}(\mathbb{P}) - H^{-}(\mathbb{P}) \tag{1.7}$$

for every path for $\mathscr{F}$.

A path $\mathbb{P}$ is called an *adiabatic path* if $\mathbb{P}$ has a standard parameterization $(\overline{V}, \overline{\theta})$ such that $\bar{\mathrm{h}}(\tau) = 0$ at all but a finite number of times τ in $[0,1]$. In view of (1.4), it is natural to study adiabatic paths by considering the initial-value problem

$$(*)\begin{cases} \dfrac{d\theta}{dV} = -\dfrac{\tilde{\lambda}(V,\theta)}{\sigma(V,\theta)}, \\ \theta(V^{\circ}) = \theta^{\circ}. \end{cases}$$

($\mathscr{F}$2) in Definition 1.1 assures us that this initial-value problem has a unique solution $\theta = \hat{\theta}(V)$ whose graph extends in both directions to meet the boundary of Σ. We call this curve the *adiabat* for $\mathscr{F}$ through $(V^{\circ}, \theta^{\circ})$ (Figure 4). The uniqueness of solutions of $(*)$ guarantees that the adiabats through $(V^{\circ}, \theta^{\circ})$ and $(\tilde{V}^{\circ}, \tilde{\theta}^{\circ})$ coincide or are disjoint curves. Consequently, the adiabats for $\mathscr{F}$ form a simple covering of Σ and can be used as a family of coordinate curves for Σ (although we will not have occasion to do so here). All of the adiabats for $\mathscr{F}$ have positive slope if $\tilde{\lambda} < 0$ and all have

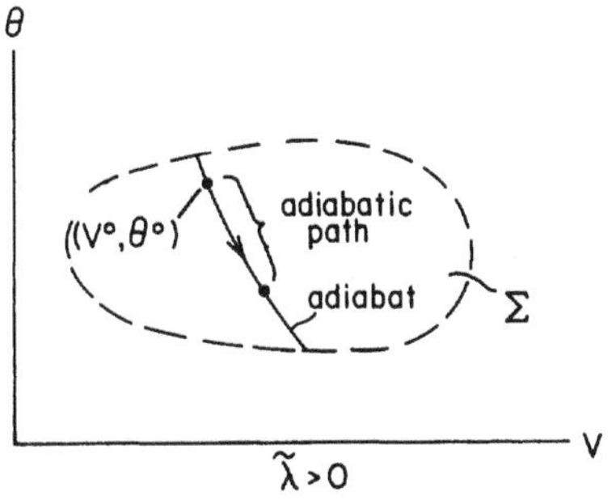

Figure 4.

negative slope if $\tilde{\lambda} > 0$. We note without proof that $\mathbb{P}$ is an adiabatic path if and only if $\mathbb{P}$ is contained in an adiabat.

An *isotherm* for $\mathscr{F}$ is the intersection of Σ with a horizontal line $\theta =$ constant, and an *isothermal path* is one which is contained in an isotherm.

Suppose $\mathbb{P}_1$ and $\mathbb{P}_2$ have standard parameterizations $(\bar{V}_1, \bar{\theta}_1)$ and $(\bar{V}_2, \bar{\theta}_2)$, respectively, such that

$$(\bar{V}_1(1), \bar{\theta}_1(1)) = (\bar{V}_2(0), \bar{\theta}_2(0)), \tag{1.8}$$

i.e., the final point of $\mathbb{P}_1$ and the initial point of $\mathbb{P}_2$ coincide. The path $\mathbb{P}_2 * \mathbb{P}_1$ is then defined by means of the standard parameterization

$$(\bar{V}(\tau), \bar{\theta}(\tau)) := \begin{cases} (\bar{V}_1(2\tau), \bar{\theta}_1(2\tau)), & 0 \leq \tau < \frac{1}{2} \\ (\bar{V}_2(2\tau - 1), \bar{\theta}_2(2\tau - 1)), & \frac{1}{2} \leq \tau \leq 1 \end{cases} \tag{1.9}$$

and is called $\mathbb{P}_1$ *followed by* $\mathbb{P}_2$. The condition (1.8) guarantees that (1.9) defines a continuous function; in fact, (1.9) provides a piecewise continuously differentiable function whose derivative vanishes at most at finitely many points in the interval [0,1]. Provided that conditions of the type (1.8) are satisfied, finite successions $\mathbb{P}_n * \cdots * \mathbb{P}_2 * \mathbb{P}_1$ of paths can be defined by repeating the above procedure. Of course, $\mathbb{P}_1 * \mathbb{P}_2$ need not be defined if $\mathbb{P}_2 * \mathbb{P}_1$ is, so the operation $*$ is not commutative, in general.

Let $\mathbb{P}$ be an arbitrary path for $\mathscr{F}$ and $(\bar{V}, \bar{\theta})$ be a standard parameterization of $\mathbb{P}$. The *reversal* $\mathbb{P}_r$ of $\mathbb{P}$ is the path determined by the standard parameterization

$$(\bar{V}_r(\tau), \bar{\theta}_r(\tau)) = (\bar{V}(1-\tau), \bar{\theta}(1-\tau)), \qquad \tau \in [0,1]. \tag{1.10}$$

A path $\mathbb{P}$ is said to be *simple* if, with the possible exception of its endpoints, no pair of states encountered along $\mathbb{P}$ are coincident. More precisely, a simple path $\mathbb{P}$ has a standard parameterization $(\bar{V}, \bar{\theta})$ such that

$$(\bar{V}(\tau_1), \bar{\theta}(\tau_1)) \neq (\bar{V}(\tau_2), \bar{\theta}(\tau_2)) \quad \text{for} \quad \tau_1 \neq \tau_2 \quad \text{and} \quad \{\tau_1, \tau_2\} \neq \{0,1\}. \tag{1.11}$$

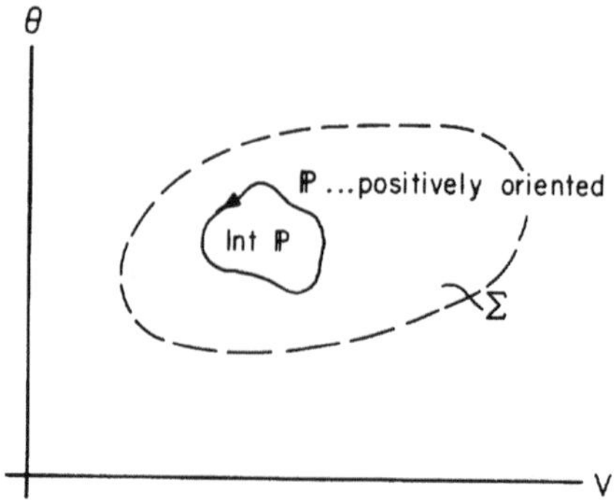

Figure 5. A simple cycle.

A path $\mathbb{P}$ is a *cycle* for $\mathscr{F}$ if its endpoints coincide, i.e.

$$(\bar{V}(1), \bar{\theta}(1)) = (\bar{V}(0), \bar{\theta}(0)) \tag{1.12}$$

for every standard parameterization of $\mathbb{P}$.

Let $\mathbb{P}$ be a *simple cycle*, i.e. a path which is both simple and a cycle. Such a path encloses a simply connected region, Int $\mathbb{P}$, the *interior* of $\mathbb{P}$ (Figure 5). A simple cycle is said to be *positively oriented* if its interior is on the left as one moves along $\mathbb{P}$ with increasing time. Otherwise, $\mathbb{P}$ is said to be *negatively oriented*.

A *Carnot path* for $\mathscr{F}$ is a finite succession of paths which are alternately adiabatic and isothermal. A *Carnot cycle* for $\mathscr{F}$ is a simple cycle which consists of two adiabatic paths and two isothermal paths. Of course, just as for adiabats, the class of Carnot cycles varies from one fluid body to another with the functions $\tilde{\lambda}$ and $\mathscr{o}$ (Figure 6). However, no matter which fluid body one considers, each state (V, θ) of $\mathscr{F}$ has a neighborhood which contains Carnot cycles for $\mathscr{F}$. Moreover, a Carnot cycle can be completely specified by giving the temperatures on the two isotherms and one point on each adiabatic path.

In the theorem which follows, we list some fundamental properties of H, H^+, H^- and W on some of the special paths just introduced.

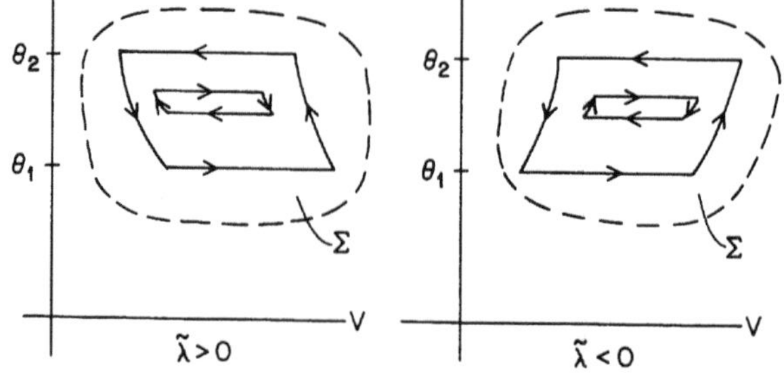

Figure 6. Typical Carnot cycles for $\mathscr{F}$.

Theorem 1.1. *Let $\mathscr{F}$ be a homogeneous fluid body. There then hold:*

(1)
$$H(\mathbb{P}_r) = -H(\mathbb{P}), \qquad H^+(\mathbb{P}_r) = H^-(\mathbb{P}),$$
$$W(\mathbb{P}_r) = -W(\mathbb{P}), \qquad H^-(\mathbb{P}_r) = H^+(\mathbb{P}),$$

for every path $\mathbb{P}$ for $\mathscr{F}$;

(2) *if $\mathbb{P}_2 * \mathbb{P}_1$ is defined and I stands for H, W, H^+ or H^-, then*
$$I(\mathbb{P}_2 * \mathbb{P}_1) = I(\mathbb{P}_2) + I(\mathbb{P}_1);$$

(3) *if $\mathbb{P}$ is a simple cycle for $\mathscr{F}$, then*
$$H(\mathbb{P}) = \mp \iint_{\operatorname{Int}\mathbb{P}} \left(\frac{\partial \tilde{\lambda}}{\partial \theta} - \frac{\partial \mathfrak{o}}{\partial V} \right) dV\, d\theta,$$
$$W(\mathbb{P}) = \mp \iint_{\operatorname{Int}\mathbb{P}} \frac{\partial p}{\partial \theta}\, dV\, d\theta,$$

where $\{\mp\}$ is taken according to whether $\mathbb{P}$ is $\left\{ \begin{matrix} \text{positively} \\ \text{negatively} \end{matrix} \right\}$ oriented.

The verification of (1)–(3) is left as an exercise. Of these properties, only (2) will carry over to more general thermodynamical systems. Property (1) depends heavily on the existence of reversals for each path for $\mathscr{F}$ and on the special behavior of line integrals on reversals; property (3) rests on the fact that Σ is a subset of the plane.

It is convenient to divide the collection of Carnot cycles for a homogeneous fluid body into two classes: (a) Carnot cycles for which heat is absorbed only at the upper temperature and emitted only at the lower temperature and (b) Carnot cycles for which heat is absorbed only at the lower temperature and emitted only at the upper temperature. A Carnot cycle of type (a) will be called a *Carnot heat engine*, while one of type (b) will be called a *Carnot refrigerator*. The operation of a Carnot heat engine can be understood if one imagines the fluid body to be inside a cylinder with a movable piston at one end. If the initial state of $\mathscr{F}$ is state ① in Figure 7, then the first adiabatic path corresponds to insulating the cylinder from its environment and compressing the fluid by moving the piston slowly toward the fixed end of the cylinder. This compression continues until $\mathscr{F}$ reaches the upper isotherm, namely at state ②. The cylinder is then placed directly in contact with a furnace whose temperature is θ_2, whereupon heat is allowed to flow into $\mathscr{F}$ while the piston moves away from the fixed wall. This expansion continues until the vertex ③ is reached, at which time the cylinder is again insulated from its environment. The fluid experiences the second adiabatic path through a continued expansion which terminates at state ④. The final path is accomplished with the cylinder in contact with a condensor at temperature θ_1 and with the fluid being compressed at that temperature until it achieves state ① once more. The operation of a Carnot refrigerator can be visualized in an analogous manner.

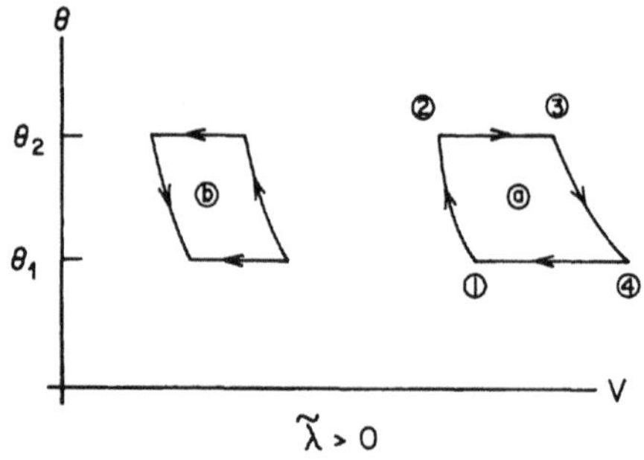

Figure 7. The two kinds of Carnot cycles.

Before we begin our discussion of the laws of thermodynamics for homogeneous fluid bodies, we introduce an important example of such a body, an ideal gas.

Definition 1.5. An *ideal gas* is a homogeneous fluid body $\mathscr{G}$ for which

$$\begin{aligned} p(V,\theta) &= \frac{R\theta}{V}, \qquad \tilde{\lambda}(V,\theta) = \frac{\lambda\theta}{V}, \\ c(V,\theta) &= \hat{c}(\theta), \qquad \Sigma = \mathbb{R}^{++} \times \mathbb{R}^{++}. \end{aligned} \tag{1.13}$$

Here, R and λ are positive numbers and $\hat{c}$ is a continuously differentiable, positive-valued function on $\mathbb{R}^{++}$.

It is easy to show that this definition is consistent with Definition 1.1, so we are justified in calling an ideal gas a homogeneous fluid body. The formula for the pressure is the classical law of Boyle and Charles to describe the behavior of gases. The formula for the latent heat $\tilde{\lambda}$ in $(1.13)_2$ is attributed to Mayer (1845) by Truesdell and Bharatha and is striking in its close resemblance to the equation of state $(1.13)_1$. The term "ideal" is appropriate to describe $\mathscr{G}$, because one need only specify two numbers R and λ and a single function $\hat{c}$ in order to obtain all of the properties of $\mathscr{G}$, throughout the *entire V-θ* quadrant, which are relevant to classical thermodynamics.

The main property of an ideal gas for our presentation is the exactness of two differentials which involve the heating and working associated with $\mathscr{G}$. This property is a consequence of the definition of an ideal gas and, thus, is a theorem in classical thermodynamics. In Section 2 we shall encounter a similar property in the context of general homogeneous fluid bodies. There it will turn out that the exactness of the corresponding differentials is equivalent to the First and Second Laws of thermodynamics for homogeneous fluid bodies.

Theorem 1.2. *Let $\mathscr{G}$ be an ideal gas and let $J = R/\lambda$. There then hold for every cycle $\mathbb{P}$ of $\mathscr{G}$:*

$$\oint_{\mathbb{P}} (J\tilde{\lambda} - p)\, dV + Jc\, d\theta = 0 \tag{1.14}$$

and

$$\oint_{\mathbb{P}} \frac{\tilde{\lambda}}{\theta}\, dV + \frac{c}{\theta}\, d\theta = 0, \tag{1.15}$$

i.e., $(\mathrm{J}\tilde{\lambda} - p)\, dV + \mathrm{J}c\, d\theta$ *and* $\tilde{\lambda}/\theta\, dV + c/\theta\, d\theta$ *are exact differentials on* Σ.

Proof. Because for every (V, θ) in Σ the functions p and $\tilde{\lambda}$ satisfy $p(V, \theta) - \mathrm{J}\tilde{\lambda}(\mathrm{V}, \theta) = 0$, it follows that

$$\oint_{\mathbb{P}} (-p + \mathrm{J}\tilde{\lambda})\, dV + \mathrm{J}c\, d\theta = \mathrm{J}\oint_{\mathbb{P}} c(\theta)\, d\theta \tag{1.16}$$

for every cycle $\mathbb{P}$. If $(\bar{V}, \bar{\theta})$ is a standard parameterization of such a cycle and if $\theta \mapsto \xi(\theta)$ is an antiderivative for c, then (1.16) implies the relations

$$\oint_{\mathbb{P}} (-p + \mathrm{J}\tilde{\lambda})\, dV + \mathrm{J}c\, d\theta = \mathrm{J}\int_0^1 c(\bar{\theta}(\tau))\bar{\theta}^{\cdot}(\tau)\, d\tau = \mathrm{J}\int_0^1 \frac{d}{d\tau}\xi(\bar{\theta}(\tau))\, d\tau$$

$$= \mathrm{J}[\xi(\bar{\theta}(1)) - \xi(\bar{\theta}(0))] = 0.$$

The last relation follows from the fact that $\bar{\theta}(1) = \bar{\theta}(0)$ for every standard parameterization of a cycle, and (1.14) has been proven. The proof of (1.15) is similar and is left as an exercise for the reader. □

The relation (1.14) can be written in the equivalent form

$$\oint_{\mathbb{P}} p\, dV = \frac{\mathrm{R}}{\lambda}\oint_{\mathbb{P}} \tilde{\lambda}\, dV + c\, d\theta,$$

i.e.,

$$W(\mathbb{P}) = \frac{\mathrm{R}}{\lambda} H(\mathbb{P}) \tag{1.17}$$

for every cycle $\mathbb{P}$ for $\mathcal{G}$. This implies that the ratio of work done to net heat gained by $\mathcal{G}$ on cycles is a constant depending only on the given parameters R and λ for $\mathcal{G}$. (1.15) asserts that the differential $\tilde{\lambda}\, dV + c\, d\theta$ associated with the heating becomes exact when multiplied by $1/\theta$. This fact is not easy to interpret directly, but it has powerful consequences when formulated for general homogeneous fluid bodies.

2. The First Law; Energy

The First and Second Laws of thermodynamics are assertions that restrict the functions W, H, H^+ and H^- for homogeneous fluid bodies. The realization that such restrictions should occur in nature is due to the great French engineer Sadi Carnot, who in 1824 published a book entitled *Reflections on the Motive Power of Heat and on Machines Suitable for Developing that Power*. Carnot there gave the essential ideas underlying the

Second Law of thermodynamics, but his work was obscured by his use of the Caloric Theory of Heat, which asserted that the net heat gained by a body along a cycle must be zero and which subsequently was proven incorrect through the beautiful experiments of Joule. It was Clausius who showed that Carnot's ideas could be recast in accordance with Joule's experiments and whose treatment of the First and Second Laws forms the basis for the versions of these laws presented in this chapter.

First Law. *There exists a positive number* J *such that, for every homogeneous fluid body* $\mathscr{F}$, *there holds*

$$W(\mathbb{P}) = JH(\mathbb{P}) \tag{2.1}$$

for every cycle $\mathbb{P}$ *for* $\mathscr{F}$.

The number J is called the *mechanical equivalent of heat*. According to the First Law, J can be determined by taking a particular homogeneous fluid body $\mathscr{F}_0$ and a particular cycle $\mathbb{P}_0$ for $\mathscr{F}_0$ and computing $W(\mathbb{P}_0)/H(\mathbb{P}_0)$. For example, (1.17) shows that the mechanical equivalent of heat equals R/λ and that R/λ has the same value for every ideal gas.

The first theorem of this section gives conditions which are equivalent to the relation (2.1).

Theorem 2.1. *Let* $\mathscr{F}$ *be a homogeneous fluid body. The following conditions are equivalent*:

(1) $W(\mathbb{P}) = JH(\mathbb{P})$ *for every cycle* $\mathbb{P}$ *for* $\mathscr{F}$;
(2) *there exists a twice continuously differentiable function* $\mathrm{E}\colon \Sigma \to \mathbb{R}$ *such that*

$$\frac{\partial \mathrm{E}}{\partial V} = \mathrm{J}\tilde{\lambda} - \mathscr{p}, \qquad \frac{\partial \mathrm{E}}{\partial \theta} = \mathrm{J}\mathscr{s}; \tag{2.2}$$

(3) *the functions* $\mathscr{p}$, $\tilde{\lambda}$, *and* $\mathscr{s}$ *satisfy*

$$\frac{\partial}{\partial \theta}(\mathrm{J}\tilde{\lambda} - \mathscr{p}) = \mathrm{J}\frac{\partial \mathscr{s}}{\partial V}. \tag{2.3}$$

This theorem tells us that the content of the First Law is equivalent to each of conditions (2) and (3). A function E satisfying (2.2) is called an (*internal*) *energy function* for $\mathscr{F}$, and each such function satisfies the relation

$$\mathrm{E}(V_2,\theta_2) - \mathrm{E}(V_1,\theta_1) = \int_{\mathbb{P}} (\mathrm{J}\tilde{\lambda} - \mathscr{p})\, dV + \mathrm{J}\mathscr{s}\, d\theta \tag{2.4}$$

for every pair of states (V_1,θ_1), $(V_2,\theta_2) \in \Sigma$ and for every path $\mathbb{P}$ from (V_1,θ_1) to (V_2,θ_2). (2.4) can also be written in the form

$$\mathrm{E}(V_2,\theta_2) - \mathrm{E}(V_1,\theta_1) = JH(\mathbb{P}) - W(\mathbb{P}). \tag{2.5}$$

On the left-hand side there appears a function of state only, while on the right a function of path. This relation is usually interpreted by saying that the quantity $JH - W$, which equals J times the heat gained by $\mathscr{F}$ *plus* the work done on $\mathscr{F}$, is "conserved" or is "path-independent." Each of the relations (2.1) and (2.5) is sometimes referred to as a statement of *conservation of energy*. It is easy to verify that two functions E and $\tilde{\mathrm{E}}$ which satisfy (2.2), (2.4) or (2.5) can differ by at most a constant through Σ. Hence, if we require in advance that internal energy functions for $\mathscr{F}$ vanish at a preassigned "standard state" for $\mathscr{F}$, then the constant must be zero and there is exactly one "normalized" internal energy function for $\mathscr{F}$. Condition (3) is significant in that it gives a condition on $\not{p}$, $\tilde{\lambda}$ and σ, the functions which define $\mathscr{F}$, which is equivalent to the relation (2.1) in the First Law. That (2.3) implies the conditions (1) and (2) rests on the fact that the state space Σ for $\mathscr{F}$, being a convex set, is simply connected. Without the simple connectivity of Σ, one can only assert that (2.3) is implied by each of conditions (1) and (2). Whether or not Σ is taken to be convex, condition (3) shows that the First Law does not permit arbitrary homogeneous fluid bodies to be present in nature. In other words, if we accept the First Law, then we must also accept the fact that materials cannot be constructed at will. As we shall see presently, the Second Law admits the same interpretation. The proof of Theorem 2.1 rests on standard results from Vector Analysis and will not be given here.

In order to avoid the presence of the symbol J in relations involving the First Law, one often assumes that units of heat and work have been chosen so as to give J the numerical value 1. In this case, (2.1) becomes

$$H(\mathbb{P}) - W(\mathbb{P}) = \oint_{\mathbb{P}} (\tilde{\lambda} - \not{p})\, dV + \sigma\, d\theta = 0 \tag{2.6}$$

for every cycle $\mathbb{P}$, and the other formulae involving J have corresponding simplifications. If one uses $\check{W}$, the work done *on* $\mathscr{F}$, rather than W, the work done by $\mathscr{F}$, then, since $\check{W} = -W$, the relation (2.1) becomes

$$JH(\mathbb{P}) + \check{W}(\mathbb{P}) = 0. \tag{2.7}$$

This form of the First Law (with or without the simplification $J = 1$) occurs frequently in the literature.

3. The Second Law; Entropy

The Second Law in the form to be given here is a condition on W and H^+ which is to hold for a special class of cycles, namely, Carnot heat engines. This restriction is in contrast to the First Law which is a condition on W and H which is to hold for *every* cycle of a homogeneous fluid body. The contrast will turn out to be illusory; we later will give a condition on arbitrary paths which is equivalent to the Second Law.

Second Law. *For every non-trivial Carnot heat engine* $\mathbb{C}$, $W(\mathbb{C})>0$. *Moreover, if* $\mathbb{C}_{\mathscr{F}_1}$ *and* $\mathbb{C}_{\mathscr{F}_2}$ *are Carnot heat engines for homogeneous fluid bodies* $\mathscr{F}_1$ *and* $\mathscr{F}_2$, *respectively, and if* $\mathbb{C}_{\mathscr{F}_1}$ *and* $\mathbb{C}_{\mathscr{F}_2}$ *have the same operating temperatures, then there holds*:

$$W(\mathbb{C}_{\mathscr{F}_1})=W(\mathbb{C}_{\mathscr{F}_2})\Rightarrow H^+(\mathbb{C}_{\mathscr{F}_1})=H^+(\mathbb{C}_{\mathscr{F}_2}). \tag{3.1}$$

By a *non-trivial* Carnot heat engine, we mean one whose isothermal segments each has non-zero length and whose operating temperatures are distinct. In the following discussion, the term "non-trivial" will be understood when we speak of Carnot heat engines. We now paraphrase Carnot's original argument for adopting the Second Law, noting in advance that our reasoning is valid whether one employs the Caloric Theory of Heat, as Carnot did, in which there holds $H(\mathbb{P})=0$ for every cycle $\mathbb{P}$ of every homogeneous fluid body, or the First Law as given by (2.1) [or its equivalent forms in Theorem 2.1]. Our argument is based on the following assertion: *no system can occur in nature which employs homogeneous fluid bodies operating through cycles in order to absorb heat at one temperature and to emit the same amount of heat at a second, higher temperature, without doing any work on its environment.* In other words, in order for heat to flow from a lower to a higher temperature, some part of a system must change its state or the system must do work (or have work done on it). As strong as this statement seems, it is quite reasonable on intuitive grounds. In fact, if a system could pump heat from a lower to a higher temperature without expending work or undergoing a change in state, our heating problems in winter would forever be solved: heat from the cold surroundings of a house could be pumped into the warm inside to compensate for the continual loss of heat through the walls and roof of the house, and this process would cost nothing in terms of expenditure of energy! We will be pragmatic and accept the idea that such a situation is impossible. Granted the validity of the italicized assertion above, our argument in support of the Second Law proceeds as follows. Suppose that there exist two homogeneous fluid bodies $\mathscr{F}_1$ and $\mathscr{F}_2$ with Carnot heat engines $\mathbb{C}_{\mathscr{F}_1}$ and $\mathbb{C}_{\mathscr{F}_2}$ satisfying

$$H^+(\mathbb{C}_{\mathscr{F}_1})>H^+(\mathbb{C}_{\mathscr{F}_2}),\qquad W(\mathbb{C}_{\mathscr{F}_1})=W(\mathbb{C}_{\mathscr{F}_2}), \tag{3.2}$$

and having the same operating temperatures $\theta_1<\theta_2$, i.e., suppose that (3.1) is false in a particular instance. For each cycle $\mathbb{P}$ of either $\mathscr{F}_1$ or $\mathscr{F}_2$ we have:

$$H^-(\mathbb{P})=H^+(\mathbb{P})-W(\mathbb{P}), \qquad \text{[in Clausius' Theory] (with J = 1)}$$

$$H^-(\mathbb{P})=H^+(\mathbb{P}), \qquad \text{[in the Caloric Theory]}$$

Using Clausius' theory together with (3.2), we obtain

$$\begin{aligned}H^-(\mathbb{C}_{\mathscr{F}_1})&=H^+(\mathbb{C}_{\mathscr{F}_1})-W(\mathbb{C}_{\mathscr{F}_1})>H^+(\mathbb{C}_{\mathscr{F}_2})-W(\mathbb{C}_{\mathscr{F}_2})\\&=H^-(\mathbb{C}_{\mathscr{F}_2})\end{aligned} \tag{3.3}$$

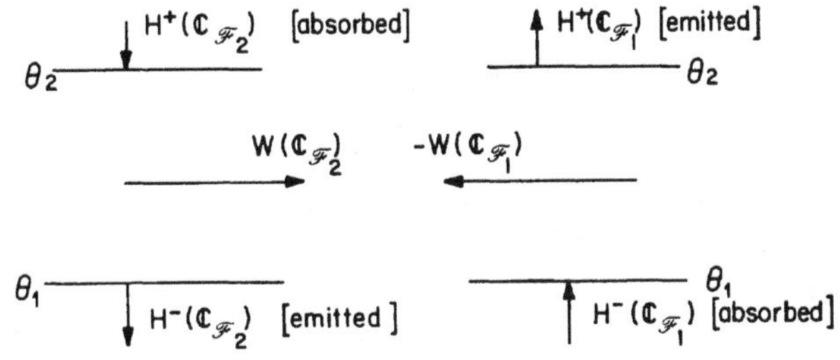

Figure 8.

and, moreover,

$$H^+(\mathbf{C}_{\mathscr{F}_1}) - H^+(\mathbf{C}_{\mathscr{F}_2}) = H^+(\mathbf{C}_{\mathscr{F}_1}) - W(\mathbf{C}_{\mathscr{F}_1}) + W(\mathbf{C}_{\mathscr{F}_2}) - H^+(\mathbf{C}_{\mathscr{F}_2})$$
$$= H^-(\mathbf{C}_{\mathscr{F}_1}) - H^-(\mathbf{C}_{\mathscr{F}_2}). \tag{3.4}$$

Consider now $\mathbf{C}_{\mathscr{F}_2}$ and $(\mathbf{C}_{\mathscr{F}_1})_r$, the reversal of $\mathbf{C}_{\mathscr{F}_1}$. The action of $(\mathbf{C}_{\mathscr{F}_1})_r$ and $\mathbf{C}_{\mathscr{F}_2}$ can be depicted schematically as in Figure 8. Carnot imagined coupling $(\mathbf{C}_{\mathscr{F}_1})_r$ and $\mathbf{C}_{\mathscr{F}_2}$ together in such a way that the interactions between these two cycles would be additive at each of the operating temperatures (see Figure 9). The positivity of $H^+(\mathbf{C}_{\mathscr{F}_1}) - H^+(\mathbf{C}_{\mathscr{F}_2})$ and of $H^-(\mathbf{C}_{\mathscr{F}_1}) - H^-(\mathbf{C}_{\mathscr{F}_2})$ as well as the vanishing of $W(\mathbf{C}_{\mathscr{F}_2}) - W(\mathbf{C}_{\mathscr{F}_1})$ follow from (3.2) through (3.4). These relations also imply that the coupled system absorbs $H^-(\mathbf{C}_{\mathscr{F}_1}) - H^-(\mathbf{C}_{\mathscr{F}_2})$ units of heat at θ_1 and emits the same quantity of heat at the higher temperature θ_2. What is more, the coupled system experiences no net change in state and does no work. Thus, the coupled system would violate our original premise, and we are forced to reject (3.2), thereby affirming (3.1) and the Second Law.

It is important to observe that the argument just given to motivate our eventual affirmation of the Second Law can be carried out using the Caloric Theory instead of the First Law. In fact, the relation $H^-(\mathbb{P}) = H^+(\mathbb{P})$ itself permits us to deduce $H^-(\mathbf{C}_{\mathscr{F}_1}) > H^-(\mathbf{C}_{\mathscr{F}_2})$ in (3.3) and $H^+(\mathbf{C}_{\mathscr{F}_1}) - H^+(\mathbf{C}_{\mathscr{F}_2}) = H^-(\mathbf{C}_{\mathscr{F}_1}) - H^-(\mathbf{C}_{\mathscr{F}_2})$ in (3.4), starting from (3.2). From that point on, the argument does not use the First Law, so one reaches the same conclusion as

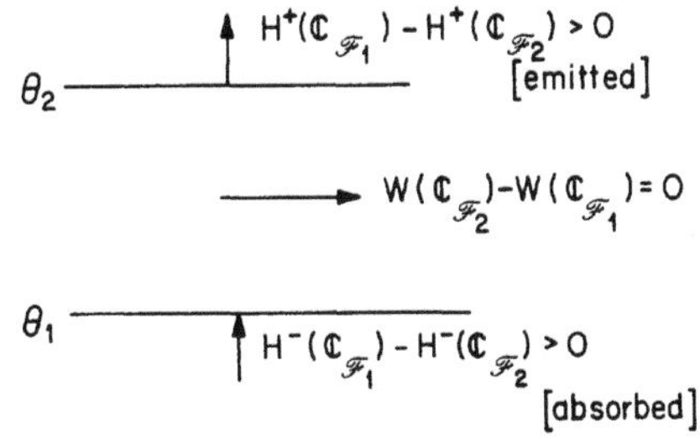

Figure 9. Coupled system.

before, namely, that (3.2) violates the original premise in our discussion. One must also keep in mind that the above argument is suggestive and heuristic in nature rather than being rigorous or precise, and we therefore cannot claim to have "proven" the Second Law on the basis of the premise above. Nevertheless, the argument is compelling in its directness and simplicity, and we shall adopt the Second Law in the form (3.1) throughout the remainder of this chapter.

The Second Law can be restated in the following equivalent way: *there exists a function* $(\theta_1, \theta_2, W) \mapsto \mathscr{A}(\theta_1, \theta_2, W)$ *such that, for every homogeneous fluid body* $\mathscr{F}$, *there holds*

$$H^+(\mathbb{C}_{\mathscr{F}}) = \mathscr{A}(\theta_1, \theta_2, W(\mathbb{C}_{\mathscr{F}})) \tag{3.5}$$

for every Carnot heat engine for $\mathscr{F}$ *operating between temperatures* $\theta_1 < \theta_2$. In other words, no matter which homogeneous fluid body is employed, the function $\mathscr{A}$ permits us to compute the heat absorbed by that body, for any Carnot heat engine, in terms of the operating temperatures and the work done by the body. Because (3.5) applies to *all* Carnot heat engines operating between θ_1 and θ_2 and to *all* homogeneous fluid bodies, we refer to $\mathscr{A}$ as a *universal* function. Moreover, if we can find a function $\mathscr{A}_{\mathscr{G}}$ such that (3.5) holds for every Carnot heat engine for a given ideal gas, then the universal function $\mathscr{A}$ must equal $\mathscr{A}_{\mathscr{G}}$. In fact, according to the Second Law, knowledge of the relationship between $\theta_1, \theta_2, W(\mathbb{C}_{\mathscr{F}})$ and $H^+(\mathbb{C}_{\mathscr{F}})$ for any *one* homogeneous fluid body $\mathscr{F}$ determines that relationship for *every* such body. These facts suggest that we attempt to determine $\mathscr{A}_{\mathscr{G}}$ from our knowledge of the functions for $\mathscr{G}$ which occur in the formulae for W and H^+: (1.1), (1.2), and (1.5). If we can find $\mathscr{A}_{\mathscr{G}}$ in this way, then we will have found $\mathscr{A}$.

In order to implement the strategy just outlined, we let θ_1, θ_2 and W be positive numbers with $\theta_1 < \theta_2$, and we attempt to construct a Carnot heat engine $\mathbb{C}_{\mathscr{G}}$ for an ideal gas $\mathscr{G}$ which operates between θ_1 and θ_2 and for which $W(\mathbb{C}_{\mathscr{G}}) = W$. To this end we consider a Carnot heat engine of the form depicted in Figure 10. Once V_1 and V_2 are given, the Carnot heat engine is determined. In fact, the adiabats through (V_1, θ_2) and (V_2, θ_2) are

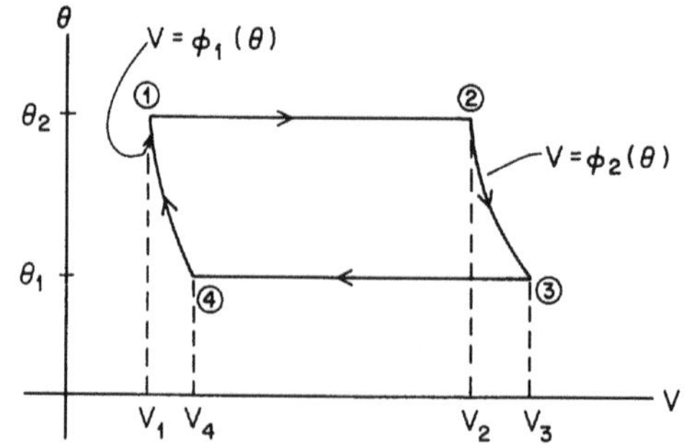

Figure 10. The cycle $\mathbb{C}_{\mathscr{G}}$.

given by

$$V=\phi_1(\theta)=V_1\exp\left[-\int_{\theta_2}^{\theta}\frac{\partial(\theta^*)}{\lambda\theta^*}d\theta^*\right], \tag{3.6}$$

$$V=\phi_2(\theta)=V_2\exp\left[-\int_{\theta_2}^{\theta}\frac{\partial(\theta^*)}{\lambda\theta^*}d\theta^*\right], \tag{3.7}$$

respectively, as can be seen by solving the initial-value problem (*) in Section 1 using the formulae in (1.13) for $\tilde{\lambda}$ and ∂ and setting $(V^\circ,\theta^\circ) = (V_1,\theta_2)$ and (V_2,θ_2), respectively. Because heat is absorbed only along the upper isothermal path, there holds

$$H^+(\mathbb{C}_{\mathscr{G}})=\int_{V_1}^{V_2}\tilde{\lambda}(V,\theta_2)\,dV=\lambda\int_{V_1}^{V_2}\frac{\theta_2}{V}dV=\lambda\theta_2\ln\frac{V_2}{V_1}. \tag{3.8}$$

The quantity $W(\mathbb{C}_{\mathscr{G}})$ is computed from (3) in Theorem 1.1 and (1.13) as follows:

$$\begin{aligned}W(\mathbb{C}_{\mathscr{G}})&=\iint_{\operatorname{Int}\mathbb{C}_{\mathscr{G}}}\frac{\mathrm{R}}{V}dV\,d\theta=\mathrm{R}\int_{\theta_1}^{\theta_2}\left(\int_{\phi_1(\theta)}^{\phi_2(\theta)}\frac{dV}{V}\right)d\theta\\&=\mathrm{R}\int_{\theta_1}^{\theta_2}\ln\frac{\phi_2(\theta)}{\phi_1(\theta)}d\theta=\mathrm{R}\int_{\theta_1}^{\theta_2}\ln\frac{V_2}{V_1}d\theta\\&=\mathrm{R}\ln\frac{V_2}{V_1}(\theta_2-\theta_1),\end{aligned} \tag{3.9}$$

where (3.6) and (3.7) have been used to compute $\ln[\phi_2(\theta)/\phi_1(\theta)]$. The numbers V_1 and V_2 can now be chosen, using (3.9), so that $W(\mathbb{C}_{\mathscr{G}})$ equals the given number W, and (3.8) and (3.9) yield

$$H^+(\mathbb{C}_{\mathscr{G}})=\frac{\theta_2}{\mathrm{J}(\theta_2-\theta_1)}W(\mathbb{C}_{\mathscr{G}}) \tag{3.10}$$

with $\mathrm{J}=\mathrm{R}/\lambda$. Because θ_1, θ_2 and W are arbitrary, this formula gives the relationship between θ_1, θ_2, $W(\mathbb{C}_{\mathscr{F}})$ and $H^+(\mathbb{C}_{\mathscr{F}})$ for every heat engine for every homogeneous fluid body, and the universal function $\mathscr{A}$ is given by

$$\mathscr{A}(\theta_1,\theta_2,W)=\frac{\theta_2}{\mathrm{J}(\theta_2-\theta_1)}W. \tag{3.11}$$

[Truesdell and Bharatha show that $\mathscr{A}$ must have the form

$$\mathscr{A}(\theta_1,\theta_2,W)=\frac{h(\theta_2)}{g(\theta_2)-g(\theta_1)}W$$

without having to employ ideal gases.]

To summarize, we have shown that the Second Law is equivalent to the assertion that

$$H^+(\mathbb{C}_{\mathscr{F}})=\frac{\theta_2}{\mathrm{J}(\theta_2-\theta_1)}W(\mathbb{C}_{\mathscr{F}}) \tag{3.12}$$

for every Carnot heat engine $\mathbb{C}_{\mathscr{F}}$ for every homogeneous fluid body $\mathscr{F}$. This formula has the immediate consequence

$$\frac{W(\mathbb{C}_{\mathscr{F}})}{JH^{+}(\mathbb{C}_{\mathscr{F}})}=1-\frac{\theta_1}{\theta_2},\tag{3.13}$$

which is *Kelvin's formula for the efficiency* $W(\mathbb{C}_{\mathscr{F}})/JH^{+}(\mathbb{C}_{\mathscr{F}})$ *of a Carnot heat engine*. Because $\theta_1<\theta_2$, Kelvin's formula tells us that this efficiency is strictly less than one and can approach one only through the use of a sequence of Carnot heat engines for which θ_1/θ_2 tends to zero. In particular, we have the striking conclusion: *no homogeneous fluid body employed as the working substance for a Carnot heat engine permits the heat absorbed to be completely converted into work*. It is instructive in this regard to rewrite (3.13), using (2.1) and (1.7), in the forms

$$H^{-}(\mathbb{C}_{\mathscr{F}})=\frac{\theta_1}{\theta_2}H^{+}(\mathbb{C}_{\mathscr{F}})\tag{3.14}$$

and

$$\theta_1=\theta_2\frac{H^{-}(\mathbb{C}_{\mathscr{F}})}{H^{+}(\mathbb{C}_{\mathscr{F}})}.\tag{3.15}$$

The formula (3.14) shows that the heat emitted is proportional to the heat absorbed and the factor of proportionality is independent of the homogeneous fluid body $\mathscr{F}$. Thus, irrespective of working substance $\mathscr{F}$, a definite portion of the heat absorbed along the upper isotherm must be emitted at the lower isotherm for a Carnot heat engine, and the Second Law can tell us how the net heat gained $H=H^{+}-H^{-}$ is distributed between heat absorbed and heat emitted for such paths. The formula (3.15) actually shows that a Carnot heat engine can be used as a thermometer whose reading will be independent of the homogeneous fluid body employed in that engine. Indeed, we need only use θ_2 as a fixed, reference temperature and calculate the temperature of an object from (3.15) by running a Carnot heat engine between the reference and unknown temperature θ_1 while measuring H^{+} and H^{-}. This calculation would be simplified by arbitrarily giving the reference temperature θ_2 the value 1° on a new scale, so that the unknown temperature on the new scale would be $H^{-}(\mathbb{C}_{\mathscr{F}})/H^{+}(\mathbb{C}_{\mathscr{F}})$. Such a temperature scale is sometimes called *absolute*, since it does not depend upon the substance used in the measuring device.

The relation (3.12) between $H^{+}(\mathbb{C}_{\mathscr{F}})$ and $W(\mathbb{C}_{\mathscr{F}})$ will now be applied to sequences of Carnot heat engines which shrink to a fixed state (V°,θ°). This procedure enables us to obtain a restriction on $\hat{\lambda}$ and $\hat{p}$ to complement the relation (2.3) obtained from the First Law. Suppose that (V°,θ°) is a state for $\mathscr{F}$ and let $\mathbb{C}_{\mathscr{F}}$ be the Carnot heat engine with (V°,θ°) as its upper right-hand vertex shown in Figure 11. It is clear that a Carnot cycle $\mathbb{C}_{\mathscr{F}}$ of this form can be constructed for $\mathscr{F}$ for θ_1 sufficiently near θ° and V_1

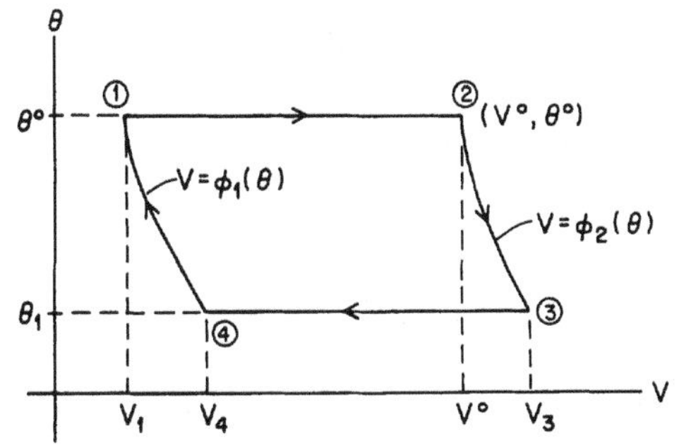

Figure 11.

sufficiently near V°. There now hold

$$H^+(\mathbf{C}_{\mathscr{F}})=\int_{V_1}^{V^\circ}\tilde{\lambda}(V,\theta^\circ)\,dV \tag{3.16}$$

and

$$\begin{aligned}W(\mathbf{C}_{\mathscr{F}})&=\iint_{\operatorname{Int}\mathbf{C}_{\mathscr{F}}}\frac{\partial p}{\partial\theta}\,dV\,d\theta\\&=\int_{\theta_1}^{\theta^\circ}\left(\int_{\phi_1(\theta)}^{\phi_2(\theta)}\frac{\partial p}{\partial\theta}(V,\theta)\,dV\right)d\theta,\end{aligned} \tag{3.17}$$

and the relation (3.12) between $H^+(\mathbf{C}_{\mathscr{F}})$ and $W(\mathbf{C}_{\mathscr{F}})$ for arbitrary Carnot heat engines yields

$$\mathrm{J}(\theta^\circ-\theta_1)\int_{V_1}^{V^\circ}\tilde{\lambda}(V,\theta^\circ)\,dV=\theta^\circ\int_{\theta_1}^{\theta^\circ}\left(\int_{\phi_1(\theta)}^{\phi_2(\theta)}\frac{\partial p}{\partial\theta}(V,\theta)\,dV\right)d\theta. \tag{3.18}$$

The relation (3.18) must hold for every choice of θ_1 sufficiently near θ° and every choice of V_1 sufficiently near V°. Therefore, we may differentiate (3.18) with respect to θ_1 to obtain

$$-\mathrm{J}\int_{V_1}^{V^\circ}\tilde{\lambda}(V,\theta^\circ)\,dV=-\theta^\circ\int_{\phi_1(\theta_1)}^{\phi_2(\theta_1)}\frac{\partial p}{\partial\theta}(V,\theta_1)\,dV,$$

and, taking the limit as θ_1 increases to θ°, we obtain

$$\int_{V_1}^{V^\circ}\tilde{\lambda}(V,\theta^\circ)\,dV=\frac{\theta^\circ}{\mathrm{J}}\int_{\phi_1(\theta^\circ)}^{\phi_2(\theta^\circ)}\frac{\partial p}{\partial\theta}(V,\theta^\circ)\,dV. \tag{3.19}$$

It follows from the construction of $\mathbf{C}_{\mathscr{F}}$ that

$$V_1=\phi_1(\theta^\circ),\qquad V^\circ=\phi_2(\theta^\circ),$$

so that (3.19) becomes

$$\int_{V_1}^{V^\circ}\tilde{\lambda}(V,\theta^\circ)\,dV=\frac{\theta^\circ}{\mathrm{J}}\int_{V_1}^{V^\circ}\frac{\partial p}{\partial\theta}(V,\theta^\circ)\,dV. \tag{3.20}$$

Because (3.20) holds for every V_1 sufficiently near V°, this relation can be differentiated with respect to V_1 to yield

$$-\tilde{\lambda}(V_1,\theta^\circ) = -\frac{\theta^\circ}{\mathrm{J}}\frac{\partial p}{\partial \theta}(V_1,\theta^\circ),$$

and, letting V_1 tend to V°, we obtain

$$\tilde{\lambda}(V^\circ,\theta^\circ) = \frac{\theta^\circ}{\mathrm{J}}\frac{\partial p}{\partial \theta}(V^\circ,\theta^\circ).$$

The state (V°,θ°) was chosen arbitrarily at the outset, and this fact permits us to write

$$\tilde{\lambda}(V,\theta) = \frac{\theta}{\mathrm{J}}\frac{\partial p}{\partial \theta}(V,\theta) \tag{3.21}$$

for *every* state (V,θ) in Σ. In other words, *the Second Law implies that the function* p *determines the function* $\tilde{\lambda}$ *through the formula* (3.21) (*called the Carnot–Clapeyron relation*). (In the derivation of (3.21), we assumed without mention that the latent heat is positive. When the latent heat is negative, a minus sign appears in front of the right-hand side of (3.16) and of (3.17), and the derivation procedes from (3.18) as above.) We are now in a position to state the following result.

Theorem 3.1. *Let* $\mathscr{F}$ *be a homogeneous fluid body. The First and Second Laws imply the formulae* (2.3) *and* (3.21) *as well as the relations*

$$\oint_{\mathbb{P}} (\mathrm{J}\tilde{\lambda} - p)\,dV + \mathrm{J}\sigma\,d\theta = 0, \tag{3.22}$$

$$\oint_{\mathbb{P}} \frac{\tilde{\lambda}(V,\theta)}{\theta}dV + \frac{\sigma(V,\theta)}{\theta}d\theta = 0, \tag{3.23}$$

which hold for every cycle $\mathbb{P}$ *for* $\mathscr{F}$.

Before giving the proof of this theorem, we note that (3.23) is equivalent to the relation

$$\int_0^1 \frac{\bar{\mathrm{h}}(\tau)}{\bar{\theta}(\tau)}d\tau = 0 \tag{3.24}$$

for every standard parameterization $(\bar{V},\bar{\theta})$ of each cycle $\mathbb{P}$ for $\mathscr{F}$. Thus, *the laws of thermodynamics imply for homogeneous fluid bodies that the heating divided by temperature has zero integral on cycles.* It is interesting to compare this result with the assertion underlying the Caloric Theory, namely, that the heating itself has zero integral on cycles. In other words, the Caloric Theory asserts that the expression

$$\tilde{\lambda}(V,\theta)\,dV + \sigma(V,\theta)\,d\theta$$

is an exact differential, while the present theory implies that

$$\frac{1}{\theta}[\tilde{\lambda}(V,\theta)\,dV+\mathscr{o}(V,\theta)\,d\theta]$$

is an exact differential. [The factor $1/\theta$ which is needed in our theory to produce an exact differential is called an *integrating factor*. Carathéodory's theory of thermodynamics is based on the interpretation of temperature as an integrating factor for the heating form. This theory was proposed in 1907 and was the first mathematical treatment of thermodynamics which received wide attention in the scientific community.]

PROOF OF THEOREM 3.1. According to Theorem 2.1, the First Law implies (3.22). Using (2.3) from that theorem, we obtain the formula

$$\frac{\partial\tilde{\lambda}}{\partial\theta}=\frac{1}{\mathrm{J}}\frac{\partial\mathscr{p}}{\partial\theta}+\frac{\partial\mathscr{o}}{\partial V},$$

which together with (3.21) yields

$$\begin{aligned}\frac{\partial}{\partial\theta}\left(\frac{\tilde{\lambda}}{\theta}\right)&=\frac{1}{\theta}\frac{\partial\tilde{\lambda}}{\partial\theta}-\frac{\tilde{\lambda}}{\theta^2}\\&=\frac{1}{\theta}\left[\frac{1}{\mathrm{J}}\frac{\partial\mathscr{p}}{\partial\theta}+\frac{\partial\mathscr{o}}{\partial V}\right]-\frac{\theta}{\mathrm{J}}\frac{\partial\mathscr{p}}{\partial\theta}\frac{1}{\theta^2}\\&=\frac{\partial}{\partial V}\left(\frac{\mathscr{o}}{\theta}\right).\end{aligned}$$

Because Σ is simply connected, this result implies that $\tilde{\lambda}/\theta\,dV+\mathscr{o}/\theta\,d\theta$ is an exact differential, so that (3.23) holds for every cycle for $\mathscr{F}$. □

In order to show that (3.22) and (3.23) express the *entire* content of the First and Second Laws, we will now prove that (3.22) and (3.23) actually imply the First and Second Laws.

Theorem 3.2. *The First and Second Laws for homogeneous fluid bodies are equivalent to the assertion that, for every homogeneous fluid body, both*

$$(\mathrm{J}\tilde{\lambda}-\mathscr{p})\,dV+\mathrm{J}\mathscr{o}\,d\theta$$

and

$$\frac{\tilde{\lambda}}{\theta}dV+\frac{\mathscr{o}}{\theta}d\theta$$

are exact differentials.

PROOF. In Theorem 3.1 we have shown that the First and Second Laws imply the exactness of the given expressions. To prove the converse, we note from Theorem 2.1 that the exactness of $(\mathrm{J}\tilde{\lambda}-\mathscr{p})\,dV+\mathrm{J}\mathscr{o}\,d\theta$ implies the First Law. If we assume also that $\tilde{\lambda}/\theta\,dV+\mathscr{o}/\theta\,d\theta$ is exact, then we can write for each Carnot heat engine $\mathbb{C}_{\mathscr{F}}$:

$$0=\oint_{\mathbb{C}_{\mathscr{F}}}\frac{\tilde{\lambda}}{\theta}dV+\frac{\mathscr{o}}{\theta}d\theta=\frac{H^{+}(\mathbb{C}_{\mathscr{F}})}{\theta_2}-\frac{H^{-}(\mathbb{C}_{\mathscr{F}})}{\theta_1}.$$

This relation along with the First Law $W(\mathbb{C}_{\mathscr{F}})=JH(\mathbb{C}_{\mathscr{F}})$ and the formula $H(\mathbb{C}_{\mathscr{F}})=H^{+}(\mathbb{C}_{\mathscr{F}})-H^{-}(\mathbb{C}_{\mathscr{F}})$ yield

$$H^{+}(\mathbb{C}_{\mathscr{F}})=\frac{\theta_2}{\theta_1}H^{-}(\mathbb{C}_{\mathscr{F}})=\frac{\theta_2}{\theta_1}\left[H^{+}(\mathbb{C}_{\mathscr{F}})-\frac{W(\mathbb{C}_{\mathscr{F}})}{J}\right],$$

or

$$H^{+}(\mathbb{C}_{\mathscr{F}})=\frac{\theta_2}{\theta_2-\theta_1}\frac{W(\mathbb{C}_{\mathscr{F}})}{J}. \tag{3.25}$$

Therefore, the exactness of the given expressions imply that the heat absorbed and the work done along each Carnot heat engine of every homogeneous fluid body are related by (3.25). Because $H^{+}(\mathbb{C}_{\mathscr{F}})$ is positive on each non-trivial Carnot heat engine, the arguments just given show that the Second Law in the original form (3.1) follows from the exactness assumed above. □

The Carnot–Clapeyron relation (3.21) can be used to obtain information about the form of *isobars*, curves of the form $p(V,\theta)=\text{const.}$, for a homogeneous fluid body. Because $\partial p/\partial V\neq 0$, the volume V can be written as a function of the temperature θ near a given point of an isobar:

$$V=\hat{V}(\theta).$$

The equation of the isobar can then be written in the form

$$p(\hat{V}(\theta),\theta)=\text{const.},$$

and, differentiating this relation with respect to θ, we obtain the relation:

$$\frac{\partial p}{\partial V}(\hat{V}(\theta),\theta)\frac{dV}{d\theta}+\frac{\partial p}{\partial \theta}(\hat{V}(\theta),\theta)=0.$$

The Carnot–Clapeyron relation then implies

$$\frac{dV}{d\theta}=-\frac{\frac{\partial p}{\partial \theta}(\hat{V}(\theta),\theta)}{\frac{\partial p}{\partial V}(\hat{V}(\theta),\theta)}=-\frac{J}{\theta}\frac{\tilde{\lambda}(\hat{V}(\theta),\theta)}{\frac{\partial p}{\partial V}(\hat{V}(\theta),\theta)} \tag{3.26}$$

on the isobar $V=\hat{V}(\theta)$. Because $\partial p/\partial V$ is negative, (3.26) shows that $dV/d\theta$ has the same sign as $\tilde{\lambda}$ along the isobar. The general forms of isobars and adiabats for a homogeneous fluid body are illustrated in Figure 12. It can be shown that the isobars and adiabats form a coordinate system for Σ which can be used in place of the V-θ system in problems where conditions of constant pressure or zero heating are to be expected. The relation (3.26)

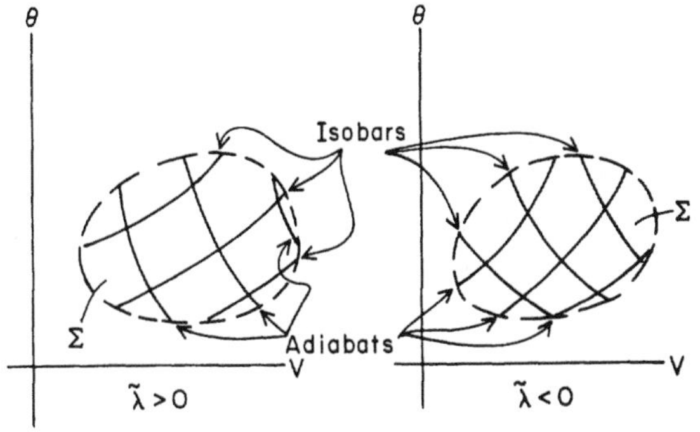

Figure 12.

also shows that, since $\tilde{\lambda}(V,\theta) \neq 0$ (($\mathscr{F}$3), Definition 1.1), $dV/d\theta \neq 0$ on an isobar in the present theory. In other words, the volume as a function of temperature on an isobar can have no maxima or minima. Since the density is proportional to the reciprocal of the volume, it follows that the density as a function of temperature on an isobar can have no maxima or minima. Therefore, our theory does not cover the so-called "anomalous behavior" of water at atmospheric pressure and temperature 4° Centigrade. [At this point on the isobar $\not{p}(V,\theta)=1$ atm, the density of water attains a maximum value.]

We saw that the exactness of $(\mathrm{J}\tilde{\lambda} - \not{p})\, dV + \mathrm{J}\sigma\, d\theta$ implied the existence of a potential E for this expression which was unique up to an additive constant and satisfied

$$\mathrm{E}(V_2,\theta_2) - \mathrm{E}(V_1,\theta_1) = \int_{\mathbb{P}} (\mathrm{J}\tilde{\lambda} - \not{p})\, dV + \mathrm{J}\sigma\, d\theta \tag{3.27}$$

for every pair of states (V_1,θ_1) and (V_2,θ_2) and every path $\mathbb{P}$ from (V_1,θ_1) to (V_2,θ_2). Such a function E we called an internal energy function for the particular homogeneous fluid body $\mathscr{F}$ in question. In exactly the same manner, the exactness of $(\tilde{\lambda}/\theta)\, dV + (\sigma/\theta)\, d\theta$ yields a potential S on Σ with the same uniqueness property and which satisfies

$$\mathrm{S}(V_2,\theta_2) - \mathrm{S}(V_1,\theta_1) = \int_{\mathbb{P}} \frac{\tilde{\lambda}(V,\theta)}{\theta}\, dV + \frac{\sigma(V,\theta)}{\theta}\, d\theta \tag{3.28}$$

with (V_1,θ_1), (V_2,θ_2) and $\mathbb{P}$ as in (3.27). A function S which satisfies (3.28) for all such states and paths is called an *entropy function* for $\mathscr{F}$. In view of Theorem 3.2 and the relations (3.27) and (3.28), we can conclude that *the First and Second Laws are equivalent to the assertion that every homogeneous fluid body possesses both an internal energy and an entropy function.*

If a "standard state" (V°,θ°) for $\mathscr{F}$ is given in advance and it is required that each internal energy function and each entropy function vanish at (V°,θ°), then the homogeneous fluid body has exactly one internal energy function E° and one entropy function S° with these properties, and it is

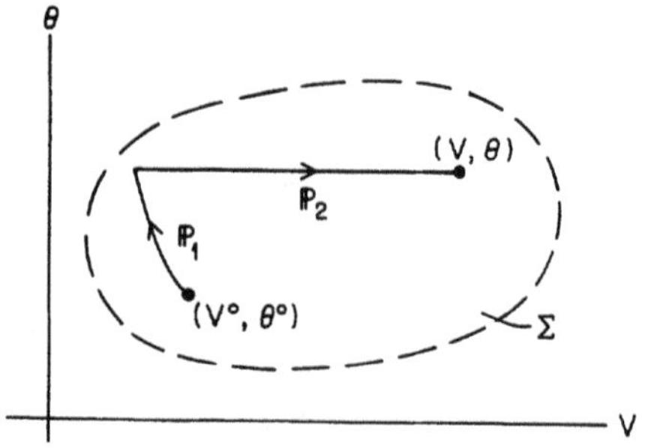

Figure 13. The path used to calculate $S^\circ(V, \theta)$.

customary to call $E^\circ(V, \theta)$ and $S^\circ(V, \theta)$ *the* internal energy and *the* entropy, respectively, of the state (V, θ). The internal energy of (V, θ) equals the number $JH(\mathbb{P}) - W(\mathbb{P})$ expressing the net heat gained by $\mathscr{F}$ minus the work done by $\mathscr{F}$ along any path from (V°, θ°) to (V, θ). It is more difficult to give an interpretation of the entropy of (V, θ), but there is a simple description available for an experiment which would permit one to calculate this quantity. Because we can write (3.28) in the form

$$S(V_2, \theta_2) - S(V_1, \theta_1) = \int_0^1 \frac{\bar{h}(\tau)}{\bar{\theta}(\tau)} d\tau, \tag{3.29}$$

it follows that S° is constant on each adiabat of $\mathscr{F}$ and

$$S^\circ(V_2, \theta) - S^\circ(V_1, \theta) = \frac{1}{\theta} H(\mathbb{P}) \tag{3.30}$$

for each isothermal path $\mathbb{P}$ at temperature θ. Thus, to find the entropy of (V, θ), it would suffice to join (V°, θ°) to (V, θ) by a path $\mathbb{P} = \mathbb{P}_2 * \mathbb{P}_1$, with $\mathbb{P}_2$ isothermal and $\mathbb{P}_1$ adiabatic, and to measure the net heat gained by $\mathscr{F}$ along $\mathbb{P}_2$ (Figure 13). From (3.29) and (3.30) one concludes that

$$S^\circ(V, \theta) = \frac{1}{\theta} H(\mathbb{P}_2),$$

i.e., the entropy of (V, θ) is the net heat gained along $\mathbb{P}_2$ divided by θ.

4. A General Efficiency Estimate

The Second Law permitted us to derive Kelvin's formula (3.13) for the efficiency of a Carnot heat engine. Let us now use (3.23), which applies to cycles which need not be Carnot heat engines, to obtain an estimate for the efficiency of a large class of cycles. The cycles which enter into this discussion are of the form $\mathbb{P} = \mathbb{P}_n * \mathbb{P}_{n-1} * \cdots * \mathbb{P}_1$ with the component segments $\mathbb{P}_1, \ldots, \mathbb{P}_n$ having standard parameterizations such that, for each

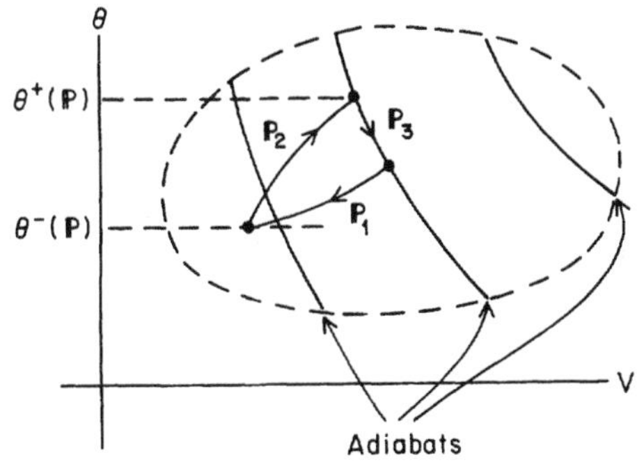

Figure 14.

$j=1,\ldots,n$, the heating $\bar{h}_j$ for $\mathbb{P}_j$ is identically zero, always positive, or always negative. For each cycle $\mathbb{P}$ of this type, we define

$$\theta^+(\mathbb{P}) = \sup\{\bar{\theta}_j(\tau) \mid \tau \in [0,1], \bar{h}_j > 0\} \tag{4.1}$$

and

$$\theta^-(\mathbb{P}) = \inf\{\bar{\theta}_j(\tau) \mid \tau \in [0,1], \bar{h}_j < 0\}; \tag{4.2}$$

$\theta^+(\mathbb{P})$ is the "largest" temperature encountered on $\mathbb{P}$ at which heat is absorbed, and $\theta^-(\mathbb{P})$ is the "smallest" temperature on $\mathbb{P}$ at which heat is emitted (Figure 14).

Theorem 4.1. *Let $\theta_1 < \theta_2$ be positive numbers, $\mathscr{F}$ a homogeneous fluid body which obeys the First and Second Laws, and $\mathbb{P}$ a cycle for $\mathscr{F}$ which is of the above form and for which*

$$H^+(\mathbb{P}) \neq 0, \qquad \theta^+(\mathbb{P}) = \theta_2, \qquad \theta^-(\mathbb{P}) = \theta_1. \tag{4.3}$$

It follows that

$$\operatorname{eff}(\mathbb{P}) := \frac{W(\mathbb{P})}{JH^+(\mathbb{P})} \leq 1 - \frac{\theta_1}{\theta_2}, \tag{4.4}$$

with equality holding if and only if the cycle $\mathbb{P}$ is a Carnot path on which heat is emitted only at temperature θ_1 and absorbed only at θ_2.

This theorem implies that Carnot heat engines are the most efficient simple cycles which emit heat only at or above a given temperature and absorb heat only at or below a second, higher temperature. Theorem 4.1 rests on (3.23), which does not mention Carnot cycles explicitly. Thus, had we adopted (3.23) from the outset as our statement of the Second Law and at first not used Carnot heat engines, we would have still obtained this theorem and discovered the special role which Carnot heat engines play in classical thermodynamics. In this light, Carnot's immediate use of these heat engines attests to his profound intuitive insight into the workings of devices for obtaining work from heat.

Proof of Theorem 4.1. From (3.23) we have

$$0=\sum_{j=1}^{n}\int_{\mathbb{P}_j}\frac{\tilde{\lambda}}{\theta}dV+\frac{\sigma}{\theta}d\theta=\sum_{j=1}^{n}\int_0^1\frac{\bar{h}_j(\tau)}{\bar{\theta}_j(\tau)}d\tau$$

$$=\sum_{\bar{h}_j>0}\int_0^1\frac{\bar{h}_j(\tau)}{\bar{\theta}_j(\tau)}d\tau+\sum_{\bar{h}_j<0}\int_0^1\frac{\bar{h}_j(\tau)}{\bar{\theta}_j(\tau)}d\tau, \qquad (4.5)$$

with the last expression displaying separately the segments of $\mathbb{P}$ on which heat is only emitted and those on which heat is only absorbed. [The adiabatic segments of $\mathbb{P}$ do not contribute to (4.5).] In each term of the first sum, the denominator $\bar{\theta}_j(\tau)$ in the integrand does not exceed $\theta^+(\mathbb{P})=\theta_2$, and, because $\bar{h}_j(\tau)$ is positive in every one of these terms, there holds

$$\sum_{\bar{h}_j>0}\int_0^1\frac{\bar{h}_j(\tau)}{\bar{\theta}_j(\tau)}d\tau\geq\frac{1}{\theta_2}\sum_{\bar{h}_j>0}\int_0^1\bar{h}_j(\tau)\,d\tau=\frac{H^+(\mathbb{P})}{\theta_2}; \qquad (4.6)_1$$

similarly,

$$\sum_{\bar{h}_j<0}\int_0^1\frac{\bar{h}_j(\tau)}{\bar{\theta}_j(\tau)}d\tau\geq\frac{1}{\theta_1}\sum_{\bar{h}_j<0}\int_0^1\bar{h}_j(\tau)\,d\tau=-\frac{H^-(\mathbb{P})}{\theta_1}, \qquad (4.6)_2$$

and these inequalities and (4.5) imply that

$$\frac{H^+(\mathbb{P})}{\theta_2}-\frac{H^-(\mathbb{P})}{\theta_1}\leq 0, \qquad (4.7)$$

which is equivalent to (4.4). It is clear from (4.5)–(4.7) that equality holds in (4.4) when the cycle $\mathbb{P}$ is a Carnot path on which heat is emitted only at θ_1 and absorbed only at θ_2. Conversely, if $\mathbb{P}$ is a cycle such that equality holds in (4.4), then (4.5) and (4.6) yield the relation

$$0=\frac{H^+(\mathbb{P})}{\theta_2}-\frac{H^-(\mathbb{P})}{\theta_1}=\sum_{\bar{h}_j>0}\int_0^1\frac{\bar{h}_j(\tau)}{\theta_2}d\tau+\sum_{\bar{h}_j<0}\int_0^1\frac{\bar{h}_j(\tau)}{\theta_1}d\tau$$

$$\leq\sum_{\bar{h}_j>0}\int_0^1\frac{\bar{h}_j(\tau)}{\bar{\theta}_j(\tau)}d\tau+\sum_{\bar{h}_j<0}\int_0^1\frac{\bar{h}_j(\tau)}{\bar{\theta}_j(\tau)}d\tau$$

$$=\oint_{\mathbb{P}}\frac{\tilde{\lambda}}{\theta}dV+\frac{\sigma}{\theta}d\theta=0.$$

It follows that " $\leq$ " in this relation must reduce to an equality which can be written in the form:

$$\sum_{\bar{h}_j>0}\int_0^1\left(\frac{1}{\bar{\theta}_j(\tau)}-\frac{1}{\theta_2}\right)\bar{h}_j(\tau)\,d\tau=\sum_{\bar{h}_j<0}\int_0^1\left(\frac{1}{\bar{\theta}_j(\tau)}-\frac{1}{\theta_1}\right)\left(-\bar{h}_j(\tau)\right)d\tau.$$

Suppose now that $\bar{\theta}_k(\bar{\tau})<\theta_2$ for one k such that $\bar{h}_k>0$. The corresponding term on the left side of the above equation would then be positive and, hence, so would be the left-hand side, itself. However, the right-hand side cannot be positive, so there must hold:

$$\bar{\mathrm{h}}_j > 0 \Rightarrow \bar{\theta}_j(\tau) \equiv \theta_2. \tag{4.8}$$

A similar argument shows that

$$\bar{\mathrm{h}}_j < 0 \Rightarrow \bar{\theta}_j(\tau) \equiv \theta_1, \tag{4.9}$$

and these relations show that heat can be absorbed only at temperature θ_2 and can be emitted only at θ_1 if the efficiency of $\mathbb{P}$ equals $1-(\theta_1/\theta_2)$. This completes the proof of Theorem 4.1. □

CHAPTER II

Systems with Perfect Accessibility

1. Definition and Example

We now begin our study of modern work on the mathematical foundations of thermodynamics. This study will be substantially more abstract than our discussion of homogeneous fluid bodies, because we here seek a presentation of thermodynamics which covers a far broader class of physical systems. At the same time, we wish to preserve as much of the logical structure of classical thermodynamics as possible. To achieve these goals, we attempt to extract from classical thermodynamics its fundamental features, and the present chapter is devoted to providing abstract counterparts of the following features of homogeneous fluid bodies: states, paths, cycles, and the expressions for work and heat.

In this section we introduce and illustrate the notion of a "system with perfect accessibility." The states of such a system are elements of a given set, the state space of the system, and are not required to have any further properties (such as being ordered pairs of positive numbers, for example). The processes of a system with perfect accessibility are described as ordered pairs consisting of an initial state and a "process generator"; the process generator for a given process always determines a final state from the initial state. The additional structure available in many examples of systems with perfect accessibility permits one to identify not only a final state, but also intermediate states for each process and a rate of traversal of these states. Of course, it is not always possible to ignore the rate of traversal as can be done in classical thermodynamics, and it is often appropriate to interpret a process not merely as a path but as a *parameterized* path in state space. Even if one adheres strictly to the level of abstraction embodied in the definition below, one can still identify the processes which can "follow" each process. Further discussion of this notion and of the term "perfect accessibility" will be given just after the formal definition.

Definition 1.1. A *system with perfect accessibility* is a pair (Σ, Π), with Σ a set whose elements σ are called *states* and Π a set whose elements π are called *process generators*, together with two functions

$$\pi \mapsto \rho_\pi,$$
$$(\pi', \pi'') \mapsto \pi''\pi'.$$

The first function assigns to each process generator π a function ρ_π, called the (*state*) *transformation* induced by π, whose domain $\mathscr{D}(\pi)$ and range $\mathscr{R}(\pi)$ are non-empty subsets of Σ. This assignment of transformations to process generators is required to satisfy the following condition of accessibility:

(S1) for each σ in Σ, there holds

$$\Pi\sigma := \{\rho_\pi \sigma | \pi \in \Pi, \sigma \in \mathscr{D}(\pi)\} = \Sigma. \tag{1.1}$$

The second function above is defined on pairs (π'', π') of process generators for which $\mathscr{D}(\pi'') \cap \mathscr{R}(\pi')$ is non-empty and assigns to each such pair a process generator $\pi''\pi'$, called the *successive application of* π'' *and* π' or π' *followed by* π''. The assignment of the process generators $\pi''\pi'$ is required to have the following property:

(S2) if $\mathscr{D}(\pi'') \cap \mathscr{R}(\pi') \neq \varnothing$, then

$$\mathscr{D}(\pi''\pi') = \rho_{\pi'}^{-1}(\mathscr{D}(\pi'')) \tag{1.2}$$

and, for each state σ in $\mathscr{D}(\pi''\pi')$, there holds

$$\rho_{\pi''\pi'}\sigma = \rho_{\pi''}\rho_{\pi'}\sigma. \tag{1.3}$$

The set of states $\Pi\sigma$ in (1.1) is called *the set of states accessible from* σ, and (S1) is the assertion that, for every state σ, the set of states accessible from σ equals the entire "state space" Σ. For this reason, we use the attributive "perfect accessibility" in our terminology. The role of the process generators $\pi''\pi'$ is to enable us to formalize the idea of performing successive operations on a system, and (S2) provides natural properties expected of successive operations. (See Figure 15 for a representation of property (S2).)

By a *process* of (Σ, Π) we mean a pair (π, σ), with σ a state and π a process generator such that σ is in $\mathscr{D}(\pi)$. Thus, given a process generator π, we obtain a process by providing a state in the domain of the transformation induced by π. The set of all processes of (Σ, Π) is denoted by $\Pi \diamond \Sigma$, i.e.,

$$\Pi \diamond \Sigma = \{(\pi, \sigma) | \pi \in \Pi, \sigma \in \mathscr{D}(\pi)\}. \tag{1.4}$$

If (π, σ) is a process, then σ is called the *initial state* for (π, σ) and $\rho_\pi \sigma$ is

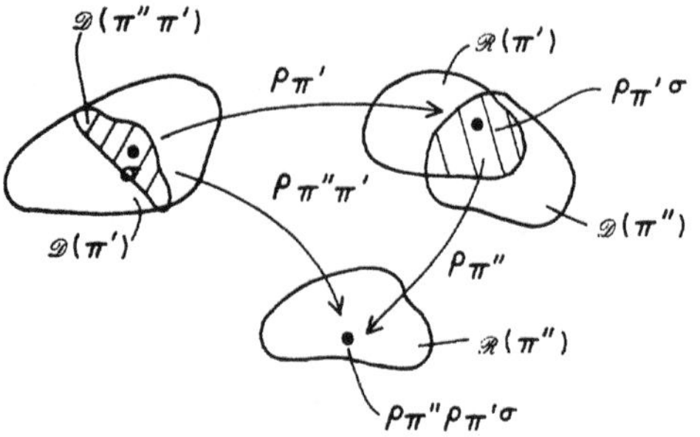

Figure 15.

called the *final state* for (π, σ). In the terminology of processes, (S1) tells us that, for each state σ, the final states for processes having initial state σ comprise the entire state space Σ. The property (S2) amounts to the following statement: if π' and π'' are such that there is a state σ' which is both the final state for a process generated by π' and the initial state for a process generated by π'', then $\pi''\pi'$ is defined and $\rho_{\pi''}\sigma'$ is the final state for a process generated by $\pi''\pi'$. In terms less precise, (S1) and (S2) tell us that every pair of states can be joined by a process, and one process can follow another when the initial state for the one is the final state for the other.

We now show how a homogeneous fluid body $\mathscr{F}$ determines a system with perfect accessibility $(\Sigma_{\mathscr{F}}, \Pi_{\mathscr{F}})$. The set $\Sigma_{\mathscr{F}}$ in the pair $(\Sigma_{\mathscr{F}}, \Pi_{\mathscr{F}})$ is taken to be the state space of $\mathscr{F}$, i.e., $\Sigma_{\mathscr{F}}$ is the open, convex subset of $\mathbb{R}^{++} \times \mathbb{R}^{++}$ in the definition of a homogeneous fluid body.[1] The set $\Pi_{\mathscr{F}}$ is defined as follows. Let $t > 0$, let π_t: $[0, t) \to \mathbb{R}^2$ be piecewise continuous, and define $\mathscr{D}(\pi_t)$ to be the set of states $\sigma = (V, \theta)$ in $\Sigma_{\mathscr{F}}$ such that the differential equation

$$(V^{\cdot}(\tau), \theta^{\cdot}(\tau)) = \pi_t(\tau), \tag{1.5}$$

has a solution $\tau \mapsto (V(\tau), \theta(\tau))$ satisfying the initial condition

$$(V(0), \theta(0)) = \sigma \tag{1.6}$$

and having trajectory $\{(V(\tau), \theta(\tau)) | \tau \in [0, t]\}$ lying entirely in Σ. In other words, σ is in $\mathscr{D}(\pi_t)$ if and only if the pair $\sigma + \int_0^\tau \pi_t(\xi)\, d\xi$ is in Σ for every τ in $[0, t]$. We now define $\Pi_{\mathscr{F}}$ to be the set of functions π_t for which $\mathscr{D}(\pi_t)$ is not empty. For each π_t in $\Pi_{\mathscr{F}}$, we define ρ_{π_t}: $\mathscr{D}(\pi_t) \to \Sigma$ by the formula

$$\rho_{\pi_t}\sigma = \sigma + \int_0^t \pi_t(\xi)\, d\xi. \tag{1.7}$$

If $\Gamma(\pi_t, \sigma)$ denotes the path (i.e., the oriented, piecewise smooth curve) determined by the parameterization $\tau \mapsto \sigma + \int_0^\tau \pi_t(\xi)\, d\xi$, $\tau \in [0, t]$, then $\rho_{\pi_t}\sigma$ is taken to be the final point of $\Gamma(\pi_t, \sigma)$. Moreover, our definition of $\mathscr{D}(\pi_t)$

[1] Chapter I, Definition 1.1.

tells us that

$$\sigma \in \mathscr{D}(\pi_t) \Leftrightarrow \Gamma(\pi_t, \sigma) \subset \Sigma_{\mathscr{F}}. \tag{1.8}$$

This is illustrated in Figure 16. It is now easy to verify (S1): if σ_1 and σ_2 are in $\Sigma_{\mathscr{F}}$ and if π_1: $[0,1] \to \mathbb{R}^2$ is defined by

$$\pi_1(\tau) = \sigma_2 - \sigma_1,$$

then $\Gamma(\pi_1, \sigma_1)$ is the line segment from σ_1 to σ_2. Because $\Sigma_{\mathscr{F}}$ is a convex set, $\Gamma(\pi_1, \sigma_1)$ is a subset of $\Sigma_{\mathscr{F}}$, and (1.8) yields

$$\sigma_1 \in \mathscr{D}(\pi_1). \tag{1.9}$$

Furthermore, σ_2 is the final point of $\Gamma(\pi_1, \sigma_1)$, so that

$$\rho_{\pi_1}\sigma_1 = \sigma_2. \tag{1.10}$$

We conclude that σ_2 is in $\Pi\sigma_1$ and, because σ_1 and σ_2 are arbitrary states in $\Sigma_{\mathscr{F}}$, the accessibility condition (S1) is satisfied.

In order to define process generators of the form $\pi''\pi'$, let $\pi_{t'}$ and $\pi_{t''}$ be in $\Pi_{\mathscr{F}}$ and define

$$\pi_{t''}\pi_{t'}(\tau) = \begin{cases} \pi_{t'}(\tau), & 0 \le \tau < t' \\ \pi_{t''}(\tau - t'), & t' \le \tau < t' + t''. \end{cases} \tag{1.11}$$

It is clear that $\pi_{t''}\pi_{t'}$ is a piecewise continuous function from $[0, t' + t'')$ into $\mathbb{R}^2$. According to (1.8), a state σ is in $\mathscr{D}(\pi_{t''}\pi_{t'})$ if and only if the curve $\Gamma(\pi_{t''}\pi_{t'}, \sigma)$ is a subset of $\Sigma_{\mathscr{F}}$. By (1.11), the segment of this parametric curve traced out for $\tau \in [0, t']$ is just $\Gamma(\pi_{t'}, \sigma)$, and, if $(V(t'), \theta(t'))$ denotes the final point of $\Gamma(\pi_{t'}, \sigma)$, the segment traced out for $\tau \in [t', t' + t'']$ is just $\Gamma(\pi_{t''}, (V(t'), \theta(t'))$. We conclude that σ is in $\mathscr{D}(\pi_{t''}\pi_{t'})$ if and only if

$$\Gamma(\pi_{t'}, \sigma) \cup \Gamma(\pi_{t''}, (V(t'), \theta(t'))) \subset \Sigma_{\mathscr{F}}. \tag{1.12}$$

Relation (1.8) shows that (1.12) is equivalent to the assertion that σ is in $\mathscr{D}(\pi_{t'})$ *and* $\rho_{\pi_{t'}}\sigma = (V(t'), \theta(t'))$ is in $\mathscr{D}(\pi_{t''})$, i.e., σ is in $\mathscr{D}(\pi_{t''}\pi_{t'})$ if and only if $\rho_{\pi_{t'}}\sigma$ is defined and is an element of $\mathscr{D}(\pi_{t''})$. It follows that $\mathscr{D}(\pi_{t''}\pi_{t'})$ is the set $\rho_{\pi_{t'}}^{-1}(\mathscr{D}(\pi_{t''}))$, and this set is non-empty if and only if $\mathscr{D}(\pi_{t''}) \cap \mathscr{R}(\pi_{t'}) \neq \varnothing$.

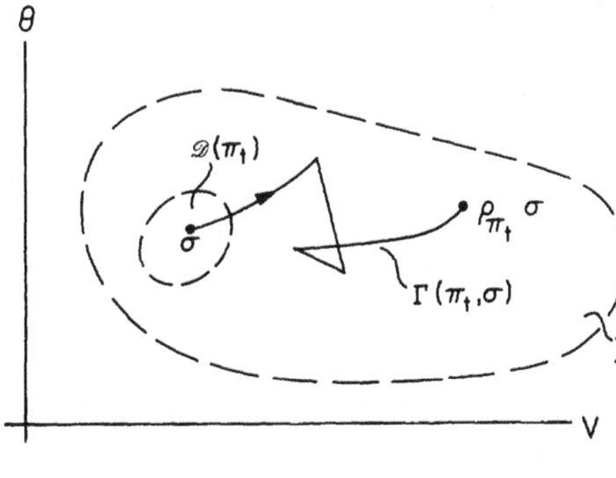

Figure 16. The path $\Gamma(\pi_t, \sigma)$ determined by (π_t, σ).

This argument verifies that $\pi_{t''}\pi_{t'}$ is in Π whenever $\mathscr{D}(\pi_{t''})\cap\mathscr{R}(\pi_{t'})$ is non-empty and that (1.2) is satisfied. Relation (1.3) follows immediately from the observation which led to (1.12): in the notation of Chapter I,

$$\Gamma(\pi_{t''}\pi_{t'},\sigma)=\Gamma(\pi_{t''},\rho_{\pi_{t'}}\sigma)*\Gamma(\pi_{t'},\sigma), \tag{1.13}$$

i.e., the path determined by $(\pi_{t''}\pi_{t'},\sigma)$ is the path determined by $(\pi_{t'},\sigma)$ followed by the path determined by $(\pi_{t''},\rho_{\pi_{t'}}\sigma)$.

The arguments just presented show that the systems described in classical thermodynamics are systems with perfect accessibility and illustrate an essential aspect of modern thermodynamics: the requirements which the abstract formulation places on systems are made explicit, so that one can check whether or not a particular system (or class of systems) meets these requirements. For each of the examples discussed in Chapters VI and VII, we do check that the requirements of Definition 1.1 are met. It should be emphasized that in applications there arise important examples which are *not* systems with perfect accessibility, and much recent effort has been devoted to organizing these examples within a more general concept of "system." Physical systems which exhibit non-trivial memory effects or which experience permanent internal changes are of the type which require a broader concept of system for their description and analysis, and an understanding of the more limited concept of systems with perfect accessibility turns out to be helpful in the study of more general systems.

We observe that the notion of "cycle" in classical thermodynamics has a counterpart for systems with perfect accessibility. We call a process (π,σ) of (Σ,Π) a *cycle* if its initial and its final state are the same, i.e., if $\rho_\pi\sigma=\sigma$. On occasion we use the terminology a *cycle through* σ and a *cycle generated by* π for a cycle of the form (π,σ), and we use the symbol $(\Pi\diamondsuit\Sigma)_{\text{cyc}}$ to denote the set of all cycles of (Σ,Π). Of course, $(\Pi\diamondsuit\Sigma)_{\text{cyc}}$ is a subset of $\Pi\diamondsuit\Sigma$, and it is an immediate consequence of (S1) that every state has a cycle through it. Moreover, if σ_1 and σ_2 are distinct states and (π',σ_1) and (π'',σ_2) are cycles, then (π',σ_1) and (π'',σ_2) are necessarily distinct (for they have different second entries), so that there are at least as many cycles as states. In particular, if Σ is an infinite set, then so is $(\Pi\diamondsuit\Sigma)_{\text{cyc}}$.

The reader may have noticed that we emphasize process generators by making them part of the prescribed structure of a system with perfect accessibility, and processes are then defined in terms of the process generators. This point of view grew on the one hand out of studies of many examples to which a general theory of the type presented here ought to apply, and it was found that process generators were at least as easy to work with as were processes, particularly in non-classical examples in which the state transformation functions ρ_π are determined through differential equations. On the other hand, the forms of the laws of thermodynamics appropriate for the broadest class of systems place natural emphasis on the initial states of processes, and our formulation in terms of process generators reflects this fact.

It is helpful to regard each process generator π as a "control" or as an "environment" in which a system in various initial states (forming the subset $\mathscr{D}(\pi)$ of Σ) can be placed. For each π, the processes of the form (π, σ) then represent all of the ways the system can evolve in the environment corresponding to π, and the set $\mathscr{R}(\pi)$ represents all of the states that can be reached in that environment.

2. Actions

The notion of an "action" is intended to translate into the language of systems with perfect accessibility the result in item (2) of Theorem 1.1, Chapter I, which asserts that the heat gained H, the work done W, the heat absorbed H^+, and the heat emitted H^- along paths of a homogeneous fluid body are additive with respect to the operation $*$. In order to achieve a generalization of this property of additivity, we observe that the above measures of heat and work are functions which assign a real number to each path of a homogeneous fluid body. Because processes for a system with perfect accessibility are the counterparts of paths, we consider functions which are defined on each process of a system (Σ, Π). In other words, we look at functions of the form

$$a: \Pi \diamond \Sigma \to \mathbb{R}.$$

Definition 2.1. An *action* for a system (Σ, Π) is a real-valued function on processes which is additive with respect to successive application, i.e., an action is a function a: $\Pi \diamond \Sigma \to \mathbb{R}$ such that

$$a(\pi''\pi', \sigma) = a(\pi', \sigma) + a(\pi'', \rho_{\pi'}\sigma), \tag{2.1}$$

for all pairs (π'', π') with $\mathscr{D}(\pi'') \cap \mathscr{R}(\pi') \neq \varnothing$ and for all states σ in $\mathscr{D}(\pi''\pi')$.

We call the number $a(\pi, \sigma)$ the *supply of a when the system undergoes the process* (π, σ), and the relation (2.1) means that the supply of a when (Σ, Π) undergoes $(\pi''\pi', \sigma)$ is the sum of the supplies of a when the system undergoes (π', σ) and $(\pi'', \rho_{\pi'}\sigma)$ successively.

Let us now show that the functions H, W, H^+ and H^- define actions for the system $(\Sigma_{\mathscr{F}}, \Pi_{\mathscr{F}})$ corresponding to a homogeneous fluid body $\mathscr{F}$. For example, for each $(\pi_t, \sigma) \in \Pi_{\mathscr{F}} \diamond \Sigma_{\mathscr{F}}$, let $\tau \mapsto (V(\tau), \theta(\tau))$ be the solution on $[0, t]$ of the initial-value problem (1.5)–(1.6) and define

$$a_H(\pi_t, \sigma) = \int_0^t [\tilde{\Lambda}(V(\tau), \theta(\tau)) V^{\cdot}(\tau) + c(V(\tau), \theta(\tau)) \theta^{\cdot}(\tau)]\, d\tau. \tag{2.2}$$

Therefore, we can write

$$a_H(\pi_t, \sigma) = \int_{\Gamma(\pi_t, \sigma)} \tilde{\lambda}(V, \theta)\, dV + \sigma(V, \theta)\, d\theta, \tag{2.3}$$

where $\Gamma(\pi_t, \sigma)$ is the path described below (1.7). In order to verify (2.1), we recall from (1.13) that

$$\Gamma(\pi_{t''}\pi_{t'}, \sigma) = \Gamma(\pi_{t''}, \rho_{\pi_{t'}}\sigma) * \Gamma(\pi_{t'}, \sigma); \tag{2.4}$$

(2.3), with π_t replaced by $\pi_{t''}\pi_{t'}$, (2.4), and the additivity of line integrals with respect to "successive application" $*$ yield:

$$\begin{aligned} a_H(\pi_{t''}\pi_{t'}, \sigma) &= \int_{\Gamma(\pi_{t''}, \rho_{\pi_{t'}}\sigma)} \tilde{\lambda}(V, \theta)\, dV + \sigma(V, \theta)\, d\theta \\ &\quad + \int_{\Gamma(\pi_{t'}, \sigma)} \tilde{\lambda}(V, \theta)\, dV + \sigma(V, \theta)\, d\theta \\ &= a_H(\pi_{t''}, \rho_{\pi_{t'}}\sigma) + a_H(\pi_{t'}, \sigma), \end{aligned}$$

which is (2.1) for a_H.

CHAPTER III

A Modern Treatment of the First Law

1. Thermodynamical Systems (1) and the First Law

As we embark on a modern treatment of the First Law, it is worth noting that the motivation and analysis presented here are brief and rather straightforward. This was also the case for the discussion in Chapter I of the First Law for homogeneous fluid bodies, and in both cases it results from the fact that, beyond a concept of system, only the notions of heat and work are needed to state the First Law. Moreover, these two interactions between a system and its environment have identical mathematical forms: both are expressed as line integrals in the classical treatment and both are actions in the present treatment.

In the next definition, we formally adjoin to a system with perfect accessibility two actions, representing heat gained and work done by the system, and call the resulting mathematical object a "thermodynamical system." (We can add the phrase "of the first type" when it is necessary to distinguish this concept from an analogous one in our presentation of the Second Law in Chapter IV.)

Definition 1.1. A *thermodynamical system* $\mathscr{S}$ is given by specifying a system with perfect accessibility $(\Sigma_{\mathscr{S}}, \Pi_{\mathscr{S}})$ and two actions $W_{\mathscr{S}}$ and $H_{\mathscr{S}}$ for $(\Sigma_{\mathscr{S}}, \Pi_{\mathscr{S}})$. The numbers $W_{\mathscr{S}}(\pi, \sigma)$ and $H_{\mathscr{S}}(\pi, \sigma)$ are called the *work done by* $\mathscr{S}$ and the *net heat gained by* $\mathscr{S}$ *in the process* (π, σ), respectively.

It is clear from the examples in Chapter II that *every homogeneous fluid body can be regarded as a thermodynamical system.*

Prior to the work of Joule in the mid-19th century, the "Caloric Theory of Heat" served as a postulate governing the behavior of systems undergoing cyclic processes. It asserted that heat, or "caloric", is conserved in the sense that all the heat which flows out of a body must return to it by the end of a cycle. In our present mathematical language, the Caloric Theory asserted that

$$(\pi, \sigma) \in (\Pi_{\mathscr{S}} \diamond \Sigma_{\mathscr{S}})_{\mathrm{cyc}} \Rightarrow H_{\mathscr{S}}(\pi, \sigma) = 0, \tag{1.1}$$

i.e., the net gain of heat in every cycle vanishes. The First Law does not contain such a strong assertion; instead, it connects $H_{\mathscr{S}}$ and $W_{\mathscr{S}}$ for those cycles on which $H_{\mathscr{S}}$ happens to vanish. In fact, the version we give here amounts to the condition that *whenever the net heat gained in a cycle vanishes, so does the work done in that cycle.*

First Law. *Let $\mathscr{S}$ be a thermodynamical system. For each cycle (π, σ) of $\mathscr{S}$, i.e., for each element of $(\Pi_{\mathscr{S}} \diamond \Sigma_{\mathscr{S}})_{cyc}$, if $H_{\mathscr{S}}(\pi, \sigma)$ vanishes then so does $W_{\mathscr{S}}(\pi, \sigma)$.*

From relation (2.1) of Chapter I, the classical First Law implies that every homogeneous fluid body obeys the present version. Our goal in the remainder of this chapter is that of showing that the First Law given here is equivalent to a statement which reduces to the classical version, namely, there is a "universal" constant which relates the work done in cycles of a thermodynamical system to the net gain of heat.

2. Products of Systems and Preservation of the First Law

Recall that Carnot's heuristic argument in support of his version of the Second Law involved combining two homogeneous fluid bodies performing Carnot cycles to form a third system for which the heat and work in the combined cycle could be obtained simply by adding the corresponding quantities for the two Carnot cycles. It is therefore not surprising that our analysis of both the First and Second Laws requires a precise description of such "combined systems." The next definition shows that, from a mathematical point of view, it is easy to provide such a description by first collecting pairs of process generators and states from two given systems with perfect accessibility to form a "product" system and then taking a work action and a heat action for the product system to be the sums of the corresponding actions for the given systems. From a physical point of view, this procedure is ambiguous, because the notion of product system admits many possible physical realizations. For example, Carnot had a particular realization in mind which depended upon the special nature of Carnot cycles and required that heat be transferred between two systems at precisely two temperatures. Clearly, this realization is too special if we are to be able to give a physical interpretation of arbitrary processes of a product system. Here we choose the following easily visualized way of interpreting the product of two systems: separate them by a "rigid, adiabatic wall" and permit them to operate, simultaneously or sequentially, without either mechanical or thermal interaction between the two systems.

Definition 2.1. Let α be an index with values 1 and 2 and let $(\Sigma_{\mathscr{S}_\alpha}, \Pi_{\mathscr{S}_\alpha})$, $W_{\mathscr{S}_\alpha}$, and $H_{\mathscr{S}_\alpha}$ be the systems with perfect accessibility and the actions specified for two thermodynamical systems $\mathscr{S}_1$ and $\mathscr{S}_2$. The *product* $\mathscr{S}_1 \times \mathscr{S}_2$ of $\mathscr{S}_1$ and $\mathscr{S}_2$ is defined through the following specifications:

$$\Sigma_{\mathscr{S}_1 \times \mathscr{S}_2} = \Sigma_{\mathscr{S}_1} \times \Sigma_{\mathscr{S}_2}, \tag{2.1}$$

$$\Pi_{\mathscr{S}_1 \times \mathscr{S}_2} = \Pi_{\mathscr{S}_1} \times \Pi_{\mathscr{S}_2}, \tag{2.2}$$

$$W_{\mathscr{S}_1 \times \mathscr{S}_2}((\pi_1, \pi_2), (\sigma_1, \sigma_2)) = W_{\mathscr{S}_1}(\pi_1, \sigma_1) + W_{\mathscr{S}_2}(\pi_2, \sigma_2), \tag{2.3}$$

$$H_{\mathscr{S}_1 \times \mathscr{S}_2}((\pi_1, \pi_2), (\sigma_1, \sigma_2)) = H_{\mathscr{S}_1}(\pi_1, \sigma_1) + H_{\mathscr{S}_2}(\pi_2, \sigma_2), \tag{2.4}$$

where the last two relations are to hold for every process (π_1, σ_1) in $\Pi_1 \diamondsuit \Sigma_1$ and (π_2, σ_2) in $\Pi_2 \diamondsuit \Sigma_2$.

According to (2.1), the states of $\mathscr{S}_1 \times \mathscr{S}_2$ are pairs of the form (σ_1, σ_2), with $\sigma_\alpha \in \Sigma_\alpha$ for $\alpha = 1, 2$, and the process generators of $\mathscr{S}_1 \times \mathscr{S}_2$ are pairs (π_1, π_2) with $\pi_\alpha \in \Pi_\alpha$, $\alpha = 1, 2$. Moreover, the pair $(\Sigma_1 \times \Sigma_2, \Pi_1 \times \Pi_2)$ can be regarded in a natural way as a system with perfect accessibility. In fact, for each process generator (π_1, π_2) we take $\rho_{(\pi_1, \pi_2)}$ to be the function from

$$\mathscr{D}(\pi_1, \pi_2) := \mathscr{D}(\pi_1) \times \mathscr{D}(\pi_2) \subset \Sigma_1 \times \Sigma_2 \tag{2.5}$$

into $\Sigma_1 \times \Sigma_2$ given by the formula

$$\rho_{(\pi_1, \pi_2)}(\sigma_1, \sigma_2) = (\rho_{\pi_1}\sigma_1, \rho_{\pi_2}\sigma_2). \tag{2.6}$$

If the set

$$\begin{aligned}\mathscr{D}((\pi_1'', \pi_2'')) \cap \mathscr{R}((\pi_1', \pi_2')) &= (\mathscr{D}(\pi_1'') \times \mathscr{D}(\pi_2'')) \cap (\mathscr{R}(\pi_1') \times \mathscr{R}(\pi_2')) \\ &= (\mathscr{D}(\pi_1'') \cap \mathscr{R}(\pi_1')) \times (\mathscr{D}(\pi_2'') \cap \mathscr{R}(\pi_2'))\end{aligned}$$

is non-empty, then we define

$$(\pi_1'', \pi_2'')(\pi_1', \pi_2') = (\pi_1''\pi_1', \pi_2''\pi_2'). \tag{2.7}$$

The relations (2.5)–(2.7) can be used to show that $(\Sigma_1 \times \Sigma_2, \Pi_1 \times \Pi_2)$ obeys conditions (S1) and (S2) in Definition 1.1 of Chapter II. Therefore,

$$(\Sigma_{\mathscr{S}_1 \times \mathscr{S}_2}, \Pi_{\mathscr{S}_1 \times \mathscr{S}_2}) = (\Sigma_{\mathscr{S}_1} \times \Sigma_{\mathscr{S}_2}, \Pi_{\mathscr{S}_1} \times \Pi_{\mathscr{S}_2})$$

is a system with perfect accessibility. It also can be verified by checking directly the additivity relation (2.1) of Chapter II that $W_{\mathscr{S}_1 \times \mathscr{S}_2}$ and $H_{\mathscr{S}_1 \times \mathscr{S}_2}$

are actions for $(\Sigma_{\mathscr{S}_1\times\mathscr{S}_2}, \Pi_{\mathscr{S}_1\times\mathscr{S}_2})$, and we conclude from Definition 1.1 of this chapter that *$\mathscr{S}_1\times\mathscr{S}_2$ has the structure of a thermodynamical system.* Whenever we speak in this chapter of *the thermodynamical system $\mathscr{S}_1\times\mathscr{S}_2$*, we shall mean the one determined by the relations (2.1)–(2.7).

Let $\mathscr{S}$ be a thermodynamical system and let $\mathscr{G}$ be an ideal gas. We say that *$\mathscr{S}$ and $\mathscr{G}$ preserve the First Law with respect to the product operation* if they satisfy the following condition:

$$\text{If } \mathscr{S} \text{ obeys the First Law, then so does } \mathscr{S}\times\mathscr{G}. \tag{2.8}$$

The condition (2.8) is fundamental for our analysis of the First Law, and the same is true for an analogous condition with respect to the Second Law. As we already remarked at the beginning of this section, a condition of this type was employed by Carnot in his study of heat engines. Here, we interpret (2.8) to mean that *the First Law applies to the independent operation of the two systems $\mathscr{S}$ and $\mathscr{G}$, provided that it applies to the system $\mathscr{S}$*. Of course, we remarked in Section 1 that every homogeneous fluid body which satisfies the classical First Law satisfies the present version of the First Law so that $\mathscr{G}$ automatically satisfies the present version. Therefore, we can regard (2.8) as a statement of the following type: if both $\mathscr{S}_1$ and $\mathscr{S}_2$ obey the First Law, then so does $\mathscr{S}_1\times\mathscr{S}_2$.

3. The First Law and Joule's Relation

The following terminology is motivated by the First Law for homogeneous fluid bodies. Let $\mathscr{S}$ be a thermodynamical system and let M be a real number. We say that *$\mathscr{S}$ satisfies Joule's relation (with factor M)* if

$$(\pi,\sigma)\in(\Pi_{\mathscr{S}}\diamond\Sigma_{\mathscr{S}})_{\text{cyc}} \Rightarrow W_{\mathscr{S}}(\pi,\sigma)=MH_{\mathscr{S}}(\pi,\sigma). \tag{3.1}$$

Thus, Joule's relation is a statement which, when applied to homogeneous fluid bodies, reduces to the classical version of the First Law.

The next result is an immediate consequence of (3.1) and the present version of the First Law. We state it here as a lemma for easy reference.

Lemma 3.1. *If a thermodynamical system $\mathscr{S}$ satisfies Joule's relation for some factor M, then $\mathscr{S}$ obeys the First Law.*

We now come to the crucial result in our analysis.

Lemma 3.2. *Let $\mathscr{S}$ be a thermodynamical system, let $\mathscr{G}$ be an ideal gas, and let R and λ be the positive numbers associated with $\mathscr{G}$ (i.e., the numbers appearing in relation* (1.13) *of Definition* 1.5, *Chapter I*). *If the product system $\mathscr{S}\times\mathscr{G}$ obeys the First Law, then $\mathscr{S}$ satisfies Joule's relation with factor R/λ.*

PROOF. We prove the contrapositive statement, i.e., if $\mathscr{S}$ fails to satisfy Joule's relation with factor R/λ, then $\mathscr{S}\times\mathscr{G}$ does not obey the First Law. Thus, we suppose that there is a cycle (π,σ) of $\mathscr{S}$ such that

$$W_{\mathscr{S}}(\pi,\sigma)-(R/\lambda)H_{\mathscr{S}}(\pi,\sigma)\neq 0. \tag{3.2}$$

Now, choose a cycle $(\pi_{\mathscr{G}},\sigma_{\mathscr{G}})$ of $\mathscr{G}$ such that

$$H_{\mathscr{G}}(\pi_{\mathscr{G}},\sigma_{\mathscr{G}})=-H_{\mathscr{S}}(\pi,\sigma); \tag{3.3}$$

this can be done by taking $(\pi_{\mathscr{G}},\sigma_{\mathscr{G}})$ to be a Carnot heat engine or refrigerator and observing that, according to relations (1.17) and (3.9) in Chapter I,

$$H_{\mathscr{G}}(\pi_{\mathscr{G}},\sigma_{\mathscr{G}})=\lambda\ln\left(\frac{V_2}{V_1}\right)(\theta_2-\theta_1),$$

so that (3.3) can be satisfied by choosing θ_1, θ_2, V_1, and V_2 appropriately. Consider next the product system $\mathscr{S}\times\mathscr{G}$ and note that, because (π,σ) and $(\pi_{\mathscr{G}},\sigma_{\mathscr{G}})$ are cycles, there holds by (2.6)

$$\rho_{(\pi,\pi_{\mathscr{G}})}(\sigma,\sigma_{\mathscr{G}})=(\rho_\pi\sigma,\rho_{\pi_{\mathscr{G}}}\sigma_{\mathscr{G}})=(\sigma,\sigma_{\mathscr{G}}).$$

Hence, $((\pi,\pi_{\mathscr{G}}),(\sigma,\sigma_{\mathscr{G}}))$ is a cycle of $\mathscr{S}\times\mathscr{G}$. Relations (2.4) and (3.3) yield

$$H_{\mathscr{S}\times\mathscr{G}}((\pi,\pi_{\mathscr{G}}),(\sigma,\sigma_{\mathscr{G}}))=H_{\mathscr{S}}(\pi,\sigma)+H_{\mathscr{G}}(\pi_{\mathscr{G}},\sigma_{\mathscr{G}})=0, \tag{3.4}$$

and (2.3), (1.17) in Chapter I, (3.3), and (3.2) imply that

$$\begin{aligned}W_{\mathscr{S}\times\mathscr{G}}((\pi,\pi_{\mathscr{G}}),(\sigma,\sigma_{\mathscr{G}}))&=W_{\mathscr{S}}(\pi,\sigma)+W_{\mathscr{G}}(\pi_{\mathscr{G}},\sigma_{\mathscr{G}})\\&=W_{\mathscr{S}}(\pi,\sigma)+\frac{R}{\lambda}H_{\mathscr{G}}(\pi_{\mathscr{G}},\sigma_{\mathscr{G}})\\&=W_{\mathscr{S}}(\pi,\sigma)-\frac{R}{\lambda}H_{\mathscr{S}}(\pi,\sigma)\\&\neq 0.\end{aligned} \tag{3.5}$$

Relations (3.4), (3.5) and the fact that $((\pi,\pi_{\mathscr{G}}),(\sigma,\sigma_{\mathscr{G}}))$ is a cycle imply that $\mathscr{S}\times\mathscr{G}$ does not obey the First Law. □

It is worth emphasizing here that the only facts from Chapter I that have been used in the preceding proof follow directly from the definition of an ideal gas. In particular, every ideal gas $\mathscr{G}$ automatically satisfies Joule's relation with factor R/λ associated with $\mathscr{G}$ and, therefore, obeys the First Law.

The two lemmas established above tell us that, if $\mathscr{S}\times\mathscr{G}$ obeys the First Law, then $\mathscr{S}$ satisfies Joule's relation and, in turn, $\mathscr{S}$ itself obeys the First Law. This pair of implications is depicted in Figure 17, which shows clearly that the First Law and Joule's relation turn out to be equivalent, if, in the language of the previous section, $\mathscr{S}$ and $\mathscr{G}$ preserve the First Law with respect to the product operation. We state this result in the following theorem.

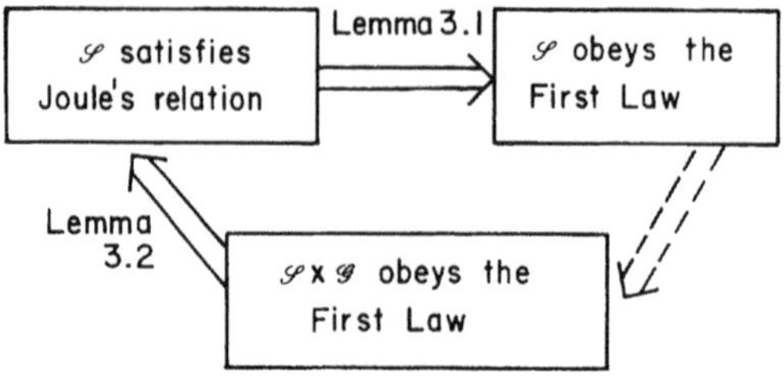

Figure 17.

Theorem 3.1. *Let $\mathscr{S}$ be a thermodynamical system, let $\mathscr{G}$ be an ideal gas, and suppose that $\mathscr{S}$ and $\mathscr{G}$ preserve the First Law with respect to the product operation. It follows that $\mathscr{S}$ obeys the First Law if and only if $\mathscr{S}$ satisfies Joule's relation with factor R/λ.*

Theorem 3.1 shows the importance of the condition (2.8): from that condition one obtains the equivalence of the First Law and Joule's relation, and Joule's relation provides information about *every* cycle of a system, not just those cycles on which the net heat gained is zero. It is interesting to ask whether (2.8) can be checked directly for the case where $\mathscr{S}$ is itself an ideal gas. The next result gives an explicit condition equivalent to (2.8) in that case. *Two ideal gases $\mathscr{G}_1$ and $\mathscr{G}_2$, with associated constants R_1, λ_1 and R_2, λ_2, preserve the First Law with respect to the product operation if and only if there holds*

$$\frac{R_1}{\lambda_1} = \frac{R_2}{\lambda_2}. \tag{3.6}$$

Proof. If $\mathscr{G}_1$ and $\mathscr{G}_2$ preserve the First Law with respect to the product operation, then Theorem 3.1, with $\mathscr{S} = \mathscr{G}_1$ and $\mathscr{G} = \mathscr{G}_2$ implies that $\mathscr{G}_1$ satisfies Joule's relation with factor R_2/λ_2. However, $\mathscr{G}_1$ satisfies Joule's relation with factor R_1/λ_1, and we conclude from (3.1) that (3.6) holds. Conversely, if (3.6) holds, then $\mathscr{G}_1$ and $\mathscr{G}_2$ will be shown to preserve the First Law if we can show from (3.6) that $\mathscr{G}_1 \times \mathscr{G}_2$ obeys the First Law. However, we have from (3.6) and the previously mentioned properties of ideal gases

$$\begin{aligned} W_{\mathscr{G}_1}(\pi_1, \sigma_1) + W_{\mathscr{G}_2}(\pi_2, \sigma_2) &= \frac{R_1}{\lambda_1} H_{\mathscr{G}_1}(\pi_1, \sigma_1) + \frac{R_2}{\lambda_2} H_{\mathscr{G}_2}(\pi_2, \sigma_2) \\ &= \frac{R_1}{\lambda_1} \left(H_{\mathscr{G}_1}(\pi_1, \sigma_1) + H_{\mathscr{G}_2}(\pi_2, \sigma_2) \right) \end{aligned}$$

for every $(\pi_1, \sigma_1) \in (\Pi_{\mathscr{G}_1} \Diamond \Sigma_{\mathscr{G}_1})_{\text{cyc}}$, $(\pi_2, \sigma_2) \in (\Pi_{\mathscr{G}_2} \Diamond \Sigma_{\mathscr{G}_2})_{\text{cyc}}$, and it follows from (2.3) and (2.4) the $\mathscr{G}_1 \times \mathscr{G}_2$ obeys the First Law. □

Relation (3.6) is of interest because it provides a method of checking whether or not two ideal gases preserve the First Law under the product operation. In fact, the ratio R/λ for an ideal gas $\mathcal{G}$ represents the ratio of the pressure to the latent heat or, more specially, the ratio of the rate of doing work to the rate of gain of heat in any isothermal process of $\mathcal{G}$. These ratios are, in principle, accessible through measurements in gases, and this gives a method of checking (2.8) when $\mathcal{S}$ is an ideal gas.

It remains to make precise the sense in which the factor R/λ in Theorem 3.1 is "universal", i.e., is the same for all elements of a collection of systems. We do this in the next theorem.

Theorem 3.2. *Let* $\mathbb{U}$ *be a collection of thermodynamical systems, and let* $\mathcal{G}$ *be an ideal gas with constants* R *and* λ. *Suppose that not only every system* $\mathcal{S}$ *in* $\mathbb{U}$ *but also all the products* $\mathcal{S} \times \mathcal{G}$ *with* $\mathcal{S}$ *in* $\mathbb{U}$ *obey the First Law. It follows that every system in* $\mathbb{U}$ *obeys Joule's relation with the same factor* R/λ.

PROOF. Because all systems $\mathcal{S} \times \mathcal{G}$ obey the First Law, it follows that (2.8) is satisfied for every $\mathcal{S}$ in $\mathbb{U}$, and we can apply Theorem 3.1 to conclude that every system $\mathcal{S}$ in $\mathbb{U}$ satisfies Joule's relation with the single factor R/λ. □

CHAPTER IV

A Modern Treatment of the Second Law

1. Introduction

In our study of the thermodynamics of homogeneous fluid bodies, we saw that the First and Second Laws implied relation (3.24) of Chapter I:

$$\int_0^1 \frac{\bar{\mathrm{h}}(\tau)}{\bar{\theta}(\tau)}\,d\tau = 0, \tag{1.1}$$

which holds on each *cycle* of a homogeneous fluid body. Conversely, granted the First Law, (1.1) was shown to imply the Second Law as well as the general efficiency estimate (4.4) in Chapter I. In other words, *the content of the Second Law is contained in relation* (1.1). In Chapter I, we could have used (1.1) from the outset had we not wanted an approach to the Second Law which had intuitive appeal and reflected the history of the subject. Indeed, use of (1.1) would offer two advantages: (1) it is mathematically more explicit than the version adopted in Chapter I; and (2) it requires only the concepts of heating and temperature. The second fact is important for the treatment of the Second Law which we present in this chapter, because it motivates an approach to the Second Law based solely on the notions of hotness and heat. In this approach neither the concept of work nor the First Law plays a role.

The pioneers in thermodynamics proposed many extensions of Carnot's version of the Second Law to more general physical systems. Lacking a mathematical framework to describe such general systems, they used "verbal" forms of the Second Law which had considerable intuitive appeal and which seemed to imply Carnot's version. The lack of precision in these verbal statements of the Second Law made it difficult to compare one with another and obscured the scope of analyses based on them. The following are the principal verbal forms of the Second Law:

1. (Kelvin) For no cyclic process of a system can heat be gained at one and only one temperature, that heat being absorbed.
2. (Kelvin–Planck) For no cyclic process can there be some heat absorbed without there ever being some heat emitted.
3. (Clausius) For no cyclic process can there be a temperature below which heat is only absorbed and above which heat is only emitted in quantity less than or equal to the amount of heat absorbed.

The third statement above is a generalization of the assertion which we used in Chapter I to motivate the Second Law for homogeneous fluid bodies in the form (3.1). A fourth verbal form of the Second Law is due to Carathéodory. From the outset, he stated this law in a precise form, and the subsequent mathematical analysis which he gave using a theory of differential forms was the first to receive wide acceptance. A verbal form of his version of the Second Law reads as follows:

4. (Carathéodory) For each state of a system, there are states arbitrarily close to it which cannot be reached from it along adiabatic paths.

Carathéodory's theory is limited to physical systems whose state spaces are of a special type, and we shall not pursue his approach in this book. Instead, we shall follow a recent treatment of the Second Law due to Serrin. Serrin uses heat and hotness as primitive concepts and a verbal form of the Second Law which is simple, precise, and which is equivalent to an integral inequality with far-reaching consequences. The exact approach which we take here differs from Serrin's in that we employ from the outset an ideal gas, just as we did in Chapter III. Serrin's approach is more general and assumes only the availability of a (possibly infinite) class of "thermometric systems" which Serrin uses to construct an "absolute" temperature scale. Here, an absolute temperature scale is identified at a certain point in the analysis as the temperature scale for the given ideal gas. Independently of and contemporaneously with the work of Serrin, Miroslav Šilhavý has given an analysis of the Second Law which in many respects is equivalent to that of Serrin. Šilhavý's treatment rests on the mathematical concept of "measure" and so goes beyond the level of knowledge assumed here. Were it not for this, Šilhavý's approach would deserve detailed discussion in this chapter.

2. Hotness Levels and Temperature Scales

The basic assumption which we make here is that we can identify for each point of a body a quantity called the "hotness" or "hotness level" of that point. In the most primitive interpretation one imagines touching the point in question and actually feeling how hot it is. Of course, in order to give the concept of hotness a quantitative meaning, one imagines using some kind of instrument with numerical readings to indicate the hotness level of a point. The choice of instruments ("thermometers") is large, but any two thermometers must agree about which of two given points is the hotter. Thus, we do not regard the actual numerical value ("temperature") given to the hotness level of a point to be fundamental; rather, it is the ordering of the hotness levels that each thermometer establishes which should be considered basic. In other words, *each pair of thermometers should establish the same ordering on the set of all hotness levels*. In practice, it is not easy to design thermometers, particularly if one wishes to take measurements during processes in which temperatures are varying rapidly or if the temperatures of points interior to a body are to be measured. These practical difficulties have caused many scientists to doubt the appropriateness of the concept of temperature or hotness in systems "outside of equilibrium." To raise such doubts about the hotness of material points without doing so also about the concept of "location of a material point" would be inconsistent on our part. In fact, similar practical difficulties arise when one attempts to measure "location," but those who reject "hotness" on account of such difficulties do *not* reject "location." Therefore, it seems necessary to accept both concepts or to reject both, and the former choice appears to be the only one consistent with the long tradition of Newtonian mechanics in which the location of a point in space is an essential primitive concept.

With such philosophical questions behind us, we turn to a precise definition of "hotness manifold," i.e. the collection of all hotness levels accessible in nature.

Definition 2.1. A *hotness manifold* is a set $\mathcal{M}$ whose elements L are called *hotness levels* together with a set $\mathcal{T}$ of functions $\varphi: \mathcal{M} \to \mathbb{R}$, called *empirical temperature scales*, satisfying:

(1) The range of φ is an open interval for each $\varphi \in \mathcal{T}$;
(2) for every $L_1, L_2 \in \mathcal{M}$ and $\varphi \in \mathcal{T}$,

$$\varphi(L_1) = \varphi(L_2) \Rightarrow L_1 = L_2;$$

(3) for every $\varphi_1, \varphi_2 \in \mathcal{T}$, $\theta \mapsto \varphi_2(\varphi_1^{-1}(\theta))$ is a continuous, strictly increasing function.

(1) asserts that each empirical temperature scale takes on all values in an open interval in $\mathbb{R}$ (which can vary from one scale to another), while (2) guarantees that each such scale establishes a one-to-one correspondence between hotness levels and real numbers in its range. By means of (3), we can establish that *the set $\mathcal{T}$ determines an order relation* $\prec$ on $\mathcal{M}$. In fact, we write:

$$L_1 \prec L_2 \Leftrightarrow \text{ there exists } \varphi \in \mathcal{T} \text{ such that } \varphi(L_1) < \varphi(L_2).$$

The relation "$\prec$" on $\mathcal{M}$ only will be meaningful if we can show that the

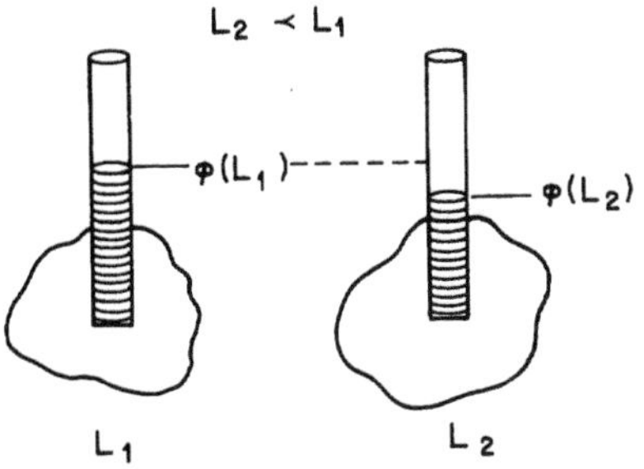

Figure 18.

situation

$$\varphi(\mathrm{L}_1) < \varphi(\mathrm{L}_2) \quad \text{and} \quad \psi(\mathrm{L}_1) \geqslant \psi(\mathrm{L}_2)$$

cannot occur for φ and ψ in $\mathscr{T}$. In fact, if $\varphi(\mathrm{L}_1) < \varphi(\mathrm{L}_2)$ and ψ is any element of $\mathscr{T}$, then letting $\theta_1 = \varphi(\mathrm{L}_1)$ and $\theta_2 = \varphi(\mathrm{L}_2)$, we have

$$\psi(L_1) = \psi(\varphi^{-1}(\varphi(\mathrm{L}_1))) = \psi(\varphi^{-1}(\theta_1)).$$

However, $\theta_1 < \theta_2$ and (3) imply that

$$\psi(\varphi^{-1}(\theta_1)) < \psi(\varphi^{-1}(\theta_2)) = \psi(\varphi^{-1}(\varphi(\mathrm{L}_2)) = \psi(\mathrm{L}_2).$$

Therefore, for every $\varphi, \psi \in \mathscr{T}$, there holds

$$\varphi(\mathrm{L}_1) < \varphi(\mathrm{L}_2) \Rightarrow \psi(\mathrm{L}_1) < \psi(\mathrm{L}_2),$$

and this shows that the relation "$\prec$" on $\mathscr{M}$ is well defined. We introduce the following terminology:

$$\mathrm{L}_1 \prec \mathrm{L}_2 \cdots \mathrm{L}_1 \textit{ is below } \mathrm{L}_2,$$
$$\mathrm{L}_1 \preccurlyeq \mathrm{L}_2 \cdots \mathrm{L}_1 \textit{ is at or below } \mathrm{L}_2.$$

Physically, one thinks of each empirical temperature scale φ as the mechanism of a thermometer which assigns a number $\varphi(\mathrm{L})$ [the temperature for L according to the scale φ] to each hotness level L. The orderings "$\prec$" and "$\preccurlyeq$" on $\mathscr{M}$ can be established by taking such a thermometer and comparing its readings on bodies at given hotness levels (Figure 18). Henceforth in this chapter, we assume given a hotness manifold $\mathscr{M}$.

3. The Accumulation Function for an Ideal Gas

We specify an ideal gas here by giving not only the state space $\Sigma_{\mathscr{G}} = \mathbb{R}^{++} \times \mathbb{R}^{++}$ and the functions $\mathscr{p}$, $\tilde{\lambda}$ and $\mathscr{c}$ as in Chapter I,

$$\mathscr{p}(V,\theta) = \frac{R\theta}{V}, \qquad \tilde{\lambda}(V,\theta) = \frac{\lambda\theta}{V} \qquad \mathscr{c}(V,\theta) = \hat{c}(\theta),$$

but also by giving an empirical temperature scale $\varphi_{\mathscr{G}}$ with range $\varphi_{\mathscr{G}} = \mathbb{R}^{++}$.

If (V, θ) is a state of $\mathscr{G}$, then we call $\varphi_{\mathscr{G}}^{-1}(\theta)$ the *hotness level of the state* (V, θ) (Figure 19). One can think of a device which determines the empirical temperature scale $\varphi_{\mathscr{G}}$: take the ideal gas at a standard pressure p_0 and, keeping its pressure constant at that value, place it in contact with an object at hotness level L and measure the volume V of the gas; the number $\varphi_{\mathscr{G}}(\mathrm{L})$ is then defined by the relation

$$\varphi_{\mathscr{G}}(\mathrm{L}) = \frac{p_0}{\mathrm{R}} V.$$

The device just described is a prototype of the gas thermometers which were used in the early stages of the development of the science of thermometry. The basis for using gas thermometers is the assumption that gases increase their volumes when placed in contact with a hotter object (always at a fixed, standard pressure).

However we might envisage the determination of the temperature scale $\varphi_{\mathscr{G}}$, the specification of this scale is now taken as part of the description of an ideal gas $\mathscr{G}$. Consider now a path $\mathbb{P}$ of $\mathscr{G}$ with standard parameterization $(\bar{V}, \bar{\theta})$ and define for each hotness level L the number $H_{\mathscr{G}}(\mathbb{P}, \mathrm{L})$ by

$$H_{\mathscr{G}}(\mathbb{P}, \mathrm{L}) := \int_{\ell(\mathbb{P}, \mathrm{L})} \bar{\mathrm{h}}(\tau)\, d\tau, \tag{3.1}$$

where $\ell(\mathbb{P}, \mathrm{L})$ is the set of times τ in $[0,1]$ at which the temperature $\bar{\theta}(\tau)$ does not exceed $\varphi_{\mathscr{G}}(\mathrm{L})$:

$$\ell(\mathbb{P}, \mathrm{L}) := \{\tau \in [0,1] \mid \bar{\theta}(\tau) \leqslant \varphi_{\mathscr{G}}(\mathrm{L})\}. \tag{3.2}$$

The integral (3.1) can be thought of as a line integral along those portions of $\mathbb{P}$ which lie on or below the horizontal line $\theta = \varphi_{\mathscr{G}}(\mathrm{L})$ in $\Sigma_{\mathscr{G}}$, and $H_{\mathscr{G}}(\mathbb{P}, \mathrm{L})$ is called the *net heat gained along* $\mathbb{P}$ *at or below* L. In order to assure that $\ell(\mathbb{P}, \mathrm{L})$ is not too complicated, as might be the case if $\mathbb{P}$ crosses $\theta = \varphi_{\mathscr{G}}(\mathrm{L})$ infinitely often, we consider here only paths $\mathbb{P}$ for $\mathscr{G}$ which cross or enter each horizontal line $\theta =$ constant at most finitely many times and have at most finitely many horizontal segments. Thus, we exclude paths of the form shown in Figure 20. It is simpler in the analysis which follows to work with

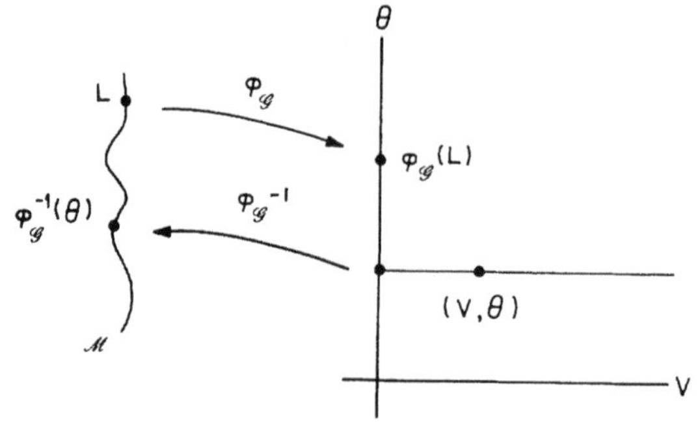

Figure 19.

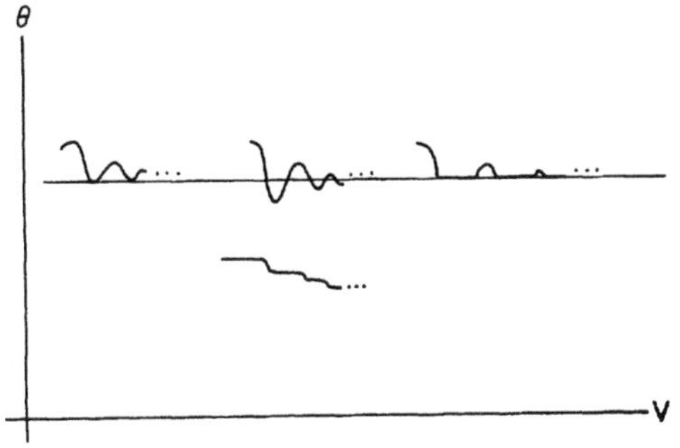

Figure 20.

the quantity

$$\hat{H}_{\mathscr{G}}(\mathbb{P},\theta):=\int_{\{\tau|\bar{\theta}(\tau)\leqslant\theta\}}\bar{\mathrm{h}}(\tau)\,d\tau=H_{\mathscr{G}}(\mathbb{P},\varphi_{\mathscr{G}}^{-1}(\theta)),\tag{3.3}$$

the *net heat gained along* $\mathbb{P}$ *at or below temperature* θ.

Properties of $\hat{H}_{\mathscr{G}}$:

1. For every path $\mathbb{P}$ for $\mathscr{G}$, the function $\theta\mapsto\hat{H}_{\mathscr{G}}(\mathbb{P},\theta)$ is zero on an interval of the form $(0,\theta_{\min})$ and constant on an interval of the form $[\theta_{\max},\infty)$, with $\theta_{\min}\leqslant\theta_{\max}$. Moreover, this function is continuous from the right at each θ in $\mathbb{R}^{++}$ and has at most a finite number of discontinuities on $\mathbb{R}^{++}$, these being jump discontinuities.

2. Let $\mathbb{P}$ be a Carnot cycle as described in Figure 21. There holds $V_1V_3=V_2V_4$, and $\hat{H}_{\mathscr{G}}(\mathbb{P},\theta)$ is given by

$$\hat{H}_{\mathscr{G}}(\mathbb{P},\theta)=\begin{cases}0 & 0<\theta<\theta_1\\ \lambda\theta_1\ln\dfrac{V_1}{V_2}, & \theta_1\leqslant\theta<\theta_2\\ \lambda(\theta_1-\theta_2)\ln\dfrac{V_1}{V_2}, & \theta_2\leqslant\theta.\end{cases}\tag{3.4}$$

3. For every path $\mathbb{P}$ with standard parameterization $(\bar{V},\bar{\theta})$, there holds

$$\int_0^\infty\hat{H}_{\mathscr{G}}(\mathbb{P},\theta)\theta^{-2}\,d\theta=\int_0^1\frac{\bar{\mathrm{h}}(\tau)}{\bar{\theta}(\tau)}\,d\tau.\tag{3.5}$$

In particular, for every cycle of $\mathscr{G}$,

$$\int_0^\infty\hat{H}_{\mathscr{G}}(\mathbb{P},\theta)\theta^{-2}\,d\theta=0.\tag{3.6}$$

We verify properties 2 and 3 and omit the proof of 1, which is somewhat technical in nature. For the Carnot heat engine shown in Figure 21, the

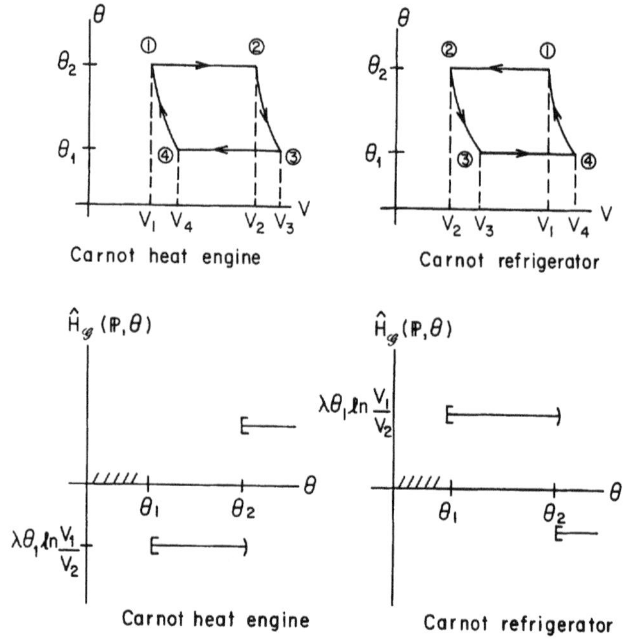

Figure 21.

adiabats are solution curves of the differential equation

$$\frac{dV}{d\theta} = -\frac{\mathscr{a}(V,\theta)}{\tilde{\lambda}(V,\theta)} = -\frac{\hat{\mathscr{a}}(\theta)}{\lambda\theta/V}$$

whose general solution is

$$V = \mathrm{D}e^{-(1/\lambda)\int(\hat{\mathscr{a}}(\theta)/\theta)\,d\theta}.$$

Therefore, the adiabats ④ ① and ② ③ have equations

$$V = V_4 e^{-(1/\lambda)\int_{\theta_1}^{\theta}(\hat{\mathscr{a}}(\xi)/\xi)\,d\xi} \text{ and } V = V_3 e^{-(1/\lambda)\int_{\theta_1}^{\theta}(\hat{\mathscr{a}}(\xi)/\xi)\,d\xi},$$

and, in particular,

$$V_1 = V_4 e^{-(1/\lambda)\int_{\theta_1}^{\theta_2}(\hat{\mathscr{a}}(\xi)/\xi)\,d\xi} \text{ and } V_2 = V_3 e^{-(1/\lambda)\int_{\theta_1}^{\theta_2}(\hat{\mathscr{a}}(\xi)/\xi)\,d\xi}.$$

The last two relations imply $V_1V_3 = V_2V_4$ for the Carnot heat engine, and the derivation of this relation for the Carnot refrigerator is analogous. For both cycles we have

$$\hat{H}_{\mathscr{G}}(\mathbb{P},\theta) = \begin{cases} 0, & \theta < \theta_1 \\ \int_{③④} \tilde{\lambda}\,dV, & \theta_1 \leqslant \theta < \theta_2 \\ \int_{③④} \tilde{\lambda}\,dV + \int_{①②} \tilde{\lambda}\,dV, & \theta_2 \leqslant \theta. \end{cases}$$

Using the formula $\tilde{\lambda}(V,\theta)=\lambda\theta/V$, we obtain

$$\int_{③④}\tilde{\lambda}\,dV=\int_{V_3}^{V_4}\frac{\lambda\theta_1}{V}dV=\lambda\theta_1\ln\frac{V_4}{V_3}=\lambda\theta_1\ln\frac{V_1}{V_2},$$

$$\int_{①②}\tilde{\lambda}\,dV=\int_{V_1}^{V_2}\frac{\lambda\theta_2}{V}dV=\lambda\theta_2\ln\frac{V_2}{V_1}=-\lambda\theta_2\ln\frac{V_1}{V_2},$$

and these relations yield the desired formula for $\hat{H}_{\mathscr{G}}(\mathbb{P},\theta)$.

The proof of 3 involves writing $\int_0^\infty\hat{H}_{\mathscr{G}}(\mathbb{P},\theta)\theta^{-2}\,d\theta$ as an iterated integral and interchanging the order of integration.

$$\int_0^\infty \hat{H}_{\mathscr{G}}(\mathbb{P},\theta)\theta^{-2}\,d\theta=\int_0^\infty\left(\int_{\{\tau|\bar{\theta}(\tau)\leqslant\theta\}}\bar{\mathrm{h}}(\tau)\,d\tau\right)\theta^{-2}\,d\theta$$
$$=\int_0^1\bar{\mathrm{h}}(\tau)\left(\int_{\bar{\theta}(\tau)}^\infty\theta^{-2}\,d\theta\right)d\tau$$
$$=\int_0^1\frac{\bar{\mathrm{h}}(\tau)}{\bar{\theta}(\tau)}d\tau.$$

[Here, each iterated integral equals the double integral over the shaded region in Figure 22.] This computation verifies (3.5), and (3.6) follows from (3.5) and relation (1.15) in Chapter I.

For our later work, we will need a result which concerns the class of functions $\theta\mapsto q(\theta)$ which are of the form

$$q(\theta)=\hat{H}_{\mathscr{G}}(\mathbb{P},\theta)$$

for some cycle $\mathbb{P}$ for $\mathscr{G}$. By a right-continuous step function on an interval $[\theta_a,\theta_b)$, we mean a function constant on subintervals $[\theta_k,\theta_{k+1})$, $k=1,2,\ldots,n$, which are disjoint and whose union is $[\theta_a,\theta_b)$.

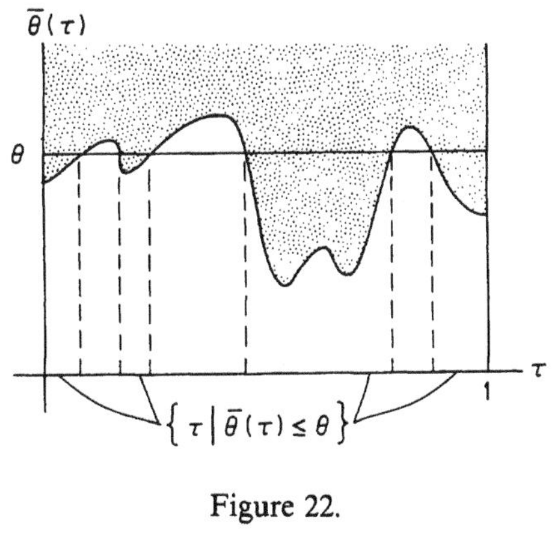

Figure 22.

Lemma. *For each right-continuous step function* $q\colon [\theta_a, \theta_b) \to \mathbb{R}$, *there is a cycle* $\mathbb{P}$ *for* $\mathcal{G}$ *such that*

$$\Theta_{\min}(\mathbb{P}) := \min\{\bar{\theta}(\tau) \mid \tau \in [0,1]\} = \theta_a \tag{3.7}$$

$$\Theta_{\max}(\mathbb{P}) := \max\{\bar{\theta}(\tau) \mid \tau \in [0,1]\} = \theta_b \tag{3.8}$$

and

$$\hat{H}_{\mathcal{G}}(\mathbb{P}, \theta) = q(\theta), \tag{3.9}$$

for every $\theta \in [\theta_a, \theta_b)$.

Instead of proving the Lemma in detail, we content ourselves with showing how $\mathbb{P}$ can be constructed when q is a step function with three values. The idea is to use Carnot heat engines and refrigerators in succession which are connected by a single adiabat. Let q be the step function shown in Figure 23a and consider the Carnot path $\mathbb{P}^* = \mathbb{P}_C * \mathbb{P}_B * \mathbb{P}_A$ in Figure 23b. $\mathbb{P}^*$ is specified by selecting an adiabat $\mathbf{A}$ and choosing as initial point for $\mathbb{P}^*$ the

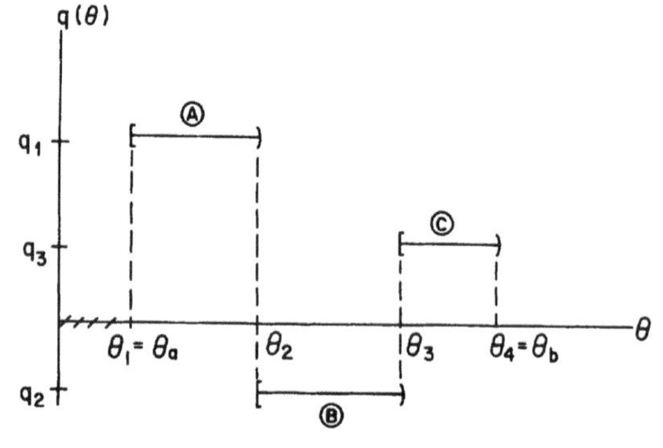

Figure 23a.

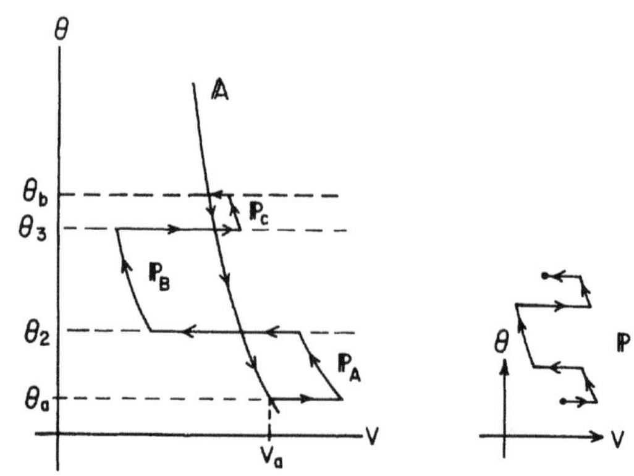

Figure 23b.

state (V_a, θ_a) where $\theta = \theta_a$ intersects $\mathbf{A}$. The path $\mathbb{P}_A$ consists of three sides of a Carnot cycle and ends at the state where $\theta = \theta_2$ intersects $\mathbf{A}$. The paths $\mathbb{P}_B$ and $\mathbb{P}_C$ are constructed in an analogous way, with $\mathbb{P}_C$ ending at the intersection of $\theta = \theta_b$ with $\mathbf{A}$. A cycle $\mathbb{P}$ is then formed by moving from the isotherm $\theta = \theta_b$ to $\theta = \theta_a$ along $\mathbf{A}$. At this stage, the horizontal dimensions of $\mathbb{P}_A$, $\mathbb{P}_B$, and $\mathbb{P}_C$ are unspecified; only the particular isotherms which form the upper and lower boundaries of $\mathbb{P}_A$, $\mathbb{P}_B$ and $\mathbb{P}_C$ have been fixed, these being the endpoints of intervals on which the given step function q is constant. The main observation needed to achieve the relation

$$\hat{H}_{\mathscr{G}}(\mathbb{P}^*, \theta) = q(\theta), \qquad \theta \in [\theta_a, \theta_b),$$

is that each of the functions $\theta \mapsto \hat{H}_{\mathscr{G}}(\mathbb{P}_A, \theta)$, $\theta \mapsto \hat{H}_{\mathscr{G}}(\mathbb{P}_B, \theta)$, and $\theta \mapsto \hat{H}_{\mathscr{G}}(\mathbb{P}_C, \theta)$ is of the form (3.4), and that $\theta \mapsto \hat{H}_{\mathscr{G}}(\mathbb{P}^*, \theta)$ is the sum of these three functions. By adjusting the lengths of the isothermal segments for $\mathbb{P}_A$, $\mathbb{P}_B$ and $\mathbb{P}_C$, the forms shown in Figure 24 for the above three functions arise. These graphs, when combined by addition, yield the graph of q on the interval $[\theta_a, \theta_b)$. [It is clear from (3.4) and these graphs that the value of

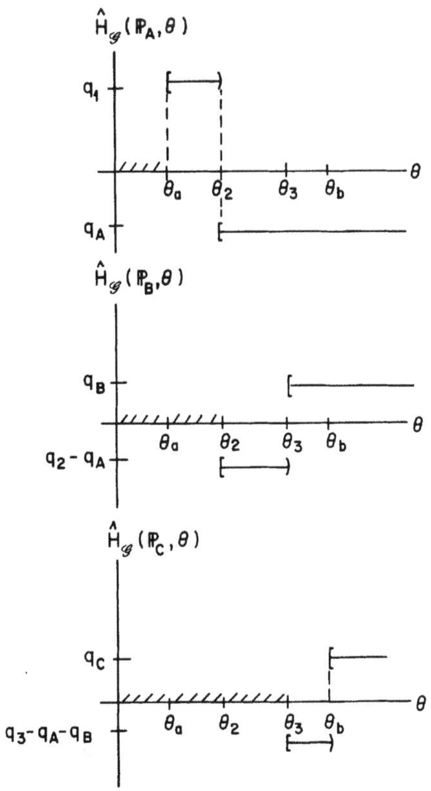

Figure 24.

$\theta \mapsto \hat{H}_{\mathscr{G}}(\mathbb{P},\theta)$ on $[\theta_b,\infty)$ cannot be adjusted at will. In fact, (3.6) can be used to show that $\hat{H}_{\mathscr{G}}(\mathbb{P},\theta_b)$ is determined by the values of $\theta \mapsto \hat{H}_{\mathscr{G}}(\mathbb{P},\theta)$ on $[\theta_a,\theta_b)$.] Note that the graphs in Figure 24 are zero on intervals $(0,\theta_a)$, $(0,\theta_2)$ and $(0,\theta_3)$ which are increasing in length.

4. Thermodynamical Systems (2) and the Second Law

Our study of the Second Law in Sections 4 through 6 of this chapter closely parallels the treatment of the First Law in Sections 1 through 3 of Chapter III. Nevertheless, the reader will find the material on the Second Law more challenging, both in terms of the motivation required to understand its modern form and the analysis needed to express it in a form suitable for applications.

We now give a definition that corresponds to Definition 1.1 in Chapter III and contains the central concept in this approach to the Second Law, that of the "accumulation function" of a thermodynamical system.

Definition 4.1. A *thermodynamical system* $\mathscr{S}$ is specified by giving a system with perfect accessibility $(\Sigma_{\mathscr{S}},\Pi_{\mathscr{S}})$ and, for each process (π,σ) of $(\Sigma_{\mathscr{S}},\Pi_{\mathscr{S}})$, a function $\mathrm{L} \mapsto H_{\mathscr{S}}(\pi,\sigma,\mathrm{L})$ from the hotness manifold $\mathscr{M}$ into the reals $\mathbb{R}$ satisfying the following three conditions:

(TS1) for every process (π,σ), there are hotness levels $\mathrm{L}' \preccurlyeq \mathrm{L}''$ such that

$$H_{\mathscr{S}}(\pi,\sigma,\mathrm{L}) = \begin{cases} 0, & \mathrm{L} \prec \mathrm{L}' \\ H_{\mathscr{S}}(\pi,\sigma,\mathrm{L}''), & \mathrm{L}'' \preccurlyeq \mathrm{L}; \end{cases}$$

(TS2) for every process (π,σ) and empirical temperature scale φ, the function $\theta \mapsto H_{\mathscr{S}}(\pi,\sigma,\varphi^{-1}(\theta))$ is right continuous with at most a finite number of discontinuities, these being jump discontinuities;

(TS3) for each hotness level L, the function $(\pi,\sigma) \mapsto H_{\mathscr{S}}(\pi,\sigma,\mathrm{L})$ is an action for $(\Sigma_{\mathscr{S}},\Pi_{\mathscr{S}})$.

The function $H_{\mathscr{S}}$ is called the *accumulation function* of $\mathscr{S}$, and $\mathrm{L} \mapsto H_{\mathscr{S}}(\pi,\sigma,\mathrm{L})$ is sometimes called the *accumulation function for* (π,σ). The value of $H_{\mathscr{S}}$ at a triple (π,σ,L) is called the *net heat gained by* $\mathscr{S}$ *in the process* (π,σ) *at or below the hotness level* L; similarly, $H_{\mathscr{S}}(\pi,\sigma,\varphi^{-1}(\theta))$ is called the *net heat gained by* $\mathscr{S}$ *in the process* (π,σ) *at or below the temperature* θ *(with respect to the empirical temperature scale* φ). According to (TS1), $\mathrm{L} \mapsto H_{\mathscr{S}}(\pi,\sigma,\mathrm{L})$ is a constant function on each of the sets $\{\mathrm{L} \in \mathscr{M} | \mathrm{L} \prec \mathrm{L}'\}$ and $\{\mathrm{L} \in \mathscr{M} | \mathrm{L}'' \preccurlyeq \mathrm{L}\}$ with value zero on the former. Moreover, even though for each process (π,σ) the hotness levels L′ and L″ in general are not uniquely determined by (π,σ), the value of $\mathrm{L} \mapsto H_{\mathscr{S}}(\pi,\sigma,\mathrm{L})$ on the latter set is independent of the choice of L″; we write $H_{\mathscr{S}}(\pi,\sigma)$ for

$H_{\mathscr{S}}(\pi, \sigma, \mathrm{L}'')$ and call it the *net heat gained by $\mathscr{S}$ in the process* (π, σ). Not only does this terminology for $H_{\mathscr{S}}(\pi, \sigma)$ agree with that in Definition 1.1 of Chapter III, but (TS3) guarantees that $(\pi, \sigma) \mapsto H_{\mathscr{S}}(\pi, \sigma)$ is an action for $(\Sigma_{\mathscr{S}}, \Pi_{\mathscr{S}})$, just as was the case in that definition. In other words, the definition of thermodynamical system in Chapter III is consistent with the present one if one identifies $H_{\mathscr{S}}(\pi, \sigma)$ in the former with $H_{\mathscr{S}}(\pi, \sigma, \mathrm{L}'')$ in the latter. Therefore, it is meaningful to consider systems $\mathscr{S}$ which are thermodynamical systems in both senses.

The present definition of thermodynamical system contains one of the fundamental requirements which we place on the physical systems covered by the theory in this chapter: *in every process of the system it is possible to identify the net heat that the system gains at or below each hotness level.* It was James Serrin who made this requirement explicit and formulated the theory which is the basis for our presentation. Of course, the idea that in thermodynamics one should distinguish between heat gained at different hotness levels can be found in Carnot's work, but no one prior to Serrin appears to have found an explicit and yet general mathematical formulation of this idea.

The accumulation function $H_{\mathscr{G}}$ of an ideal gas is given by relations (3.1) and (3.2), and property one of the function $\hat{H}_{\mathscr{G}}$ in (3.3) implies that $H_{\mathscr{G}}$ satisfies (TS1) and (TS2). We already observed in Chapter II that each homogeneous fluid body determines a system with perfect accessibility, and (3.1) together with (3.2) permit one to show that, for each hotness level L, the function $(\pi, \sigma) \mapsto H_{\mathscr{G}}(\pi, \sigma, \mathrm{L})$ is an action for $(\Sigma_{\mathscr{G}}, \Pi_{\mathscr{G}})$, i.e., (TS3) is satisfied: *each ideal gas determines a thermodynamical system in the sense of Definition* 4.1. Actually, this conclusion holds for every homogeneous fluid body $\mathscr{F}$ if in (3.1) we replace $\mathscr{G}$ by $\mathscr{F}$, $\bar{\mathrm{h}}$ by the heating for $\mathscr{F}$ along $\mathbb{P}$, and, in (3.2), we employ any empirical temperature scale. Thus, our theory applies to the homogeneous fluid bodies of classical thermodynamics, with the exclusion of paths of the type shown in Figure 20. (This exclusion can be avoided but only at the cost of employing more advanced mathematical concepts.)

When we speak of an ideal gas as a thermodynamical system, it is often convenient to interchange processes (π, σ) and the paths $\mathbb{P}$ they determine. This is usually done by interchanging the symbols $H_{\mathscr{G}}(\mathbb{P}, \mathrm{L})$ and $H_{\mathscr{G}}(\pi, \sigma, \mathrm{L})$; it causes no difficulties, because the net heat gained in a process of $\mathscr{G}$ at or below L depends only upon the path determined by the process.

If $\mathscr{S}$ is a thermodynamical system and φ is an empirical temperature scale, then the identification of $H_{\mathscr{S}}(\pi, \sigma, \varphi^{-1}(\theta))$ as the net heat gained in (π, σ) at or below temperature θ has the following consequences:

(a) if $H_{\mathscr{S}}(\pi, \sigma, \varphi^{-1}(\theta_2)) - H_{\mathscr{S}}(\pi, \sigma, \varphi^{-1}(\theta_1))$ is positive, then heat is absorbed in the process (π, σ) on the temperature interval $(\theta_1, \theta_2]$;
(b) if $H_{\mathscr{S}}(\pi, \sigma, \varphi^{-1}(\theta_2)) - H_{\mathscr{S}}(\pi, \sigma, \varphi^{-1}(\theta_1))$ is negative, then heat is emitted in the process (π, σ) on the temperature interval $(\theta_1, \theta_2]$;

(c) If $H_{\mathscr{S}}(\pi,\sigma,\varphi^{-1}(\theta_2)) - H_{\mathscr{S}}(\pi,\sigma,\varphi^{-1}(\theta_1))$ is zero, then there is no net gain of heat in (π,σ) on the interval $(\theta_1,\theta_2]$.

It follows that if $\theta \mapsto H_{\mathscr{S}}(\pi,\sigma,\varphi^{-1}(\theta))$ is $\left\{\begin{matrix}\text{decreasing}\\ \text{increasing}\end{matrix}\right\}$ on an interval $I \subset \mathbb{R}$, then heat is $\left\{\begin{matrix}\text{emitted}\\ \text{absorbed}\end{matrix}\right\}$ by $\mathscr{S}$ on each subinterval of I in the process (π,σ). Moreover, an interval I on which $\theta \mapsto H_{\mathscr{S}}(\pi,\sigma,\varphi^{-1}(\theta))$ is constant corresponds to no net gain of heat by $\mathscr{S}$ on any subinterval of I in the process (π,σ).

In those cases where processes of a system can be described by functions on time intervals, it is important to distinguish between net gain of heat on temperature intervals, as expressed by accumulation functions, and net gain of heat on time intervals. For example, consider for an ideal gas $\mathscr{G}$ an isothermal cycle (π,σ) (or $\mathbb{P}$) which has standard parameterization $(\bar{V},\bar{\theta})$ given by

$$(\bar{V}(\tau),\bar{\theta}(\tau)) = \begin{cases} (V_0 + a\tau, \theta_0), & 0 \leqslant \tau \leqslant \frac{1}{2} \\ (V_0 + a(1-\tau), \theta_0), & \frac{1}{2} < \tau \leqslant 1, \end{cases} \tag{4.1}$$

with a a positive number. The net gain of heat on the time interval $[0,\tau]$ is given by the expression

$$\begin{aligned} H(\tau) &= \int_0^\tau \bar{\lambda}(\bar{V}(\xi),\bar{\theta}(\xi))\bar{V}^{\cdot}(\xi)\,d\xi \\ &= \begin{cases} \displaystyle\int_0^\tau \frac{a\lambda\theta_0\,d\xi}{V_0 + a\xi}, & 0 \leqslant \tau \leqslant \frac{1}{2} \\ \displaystyle\int_0^{1/2} \frac{a\lambda\theta_0\,d\xi}{V_0 + a\xi} - \int_{1/2}^\tau \frac{a\lambda\theta_0\,d\xi}{V_0 + a(1-\xi)}, & \frac{1}{2} < \tau \leqslant 1 \end{cases} \\ &= \begin{cases} \lambda\theta_0 \ln\dfrac{V_0 + a\tau}{V_0}, & 0 \leqslant \tau \leqslant \frac{1}{2} \\ \lambda\theta_0 \ln\dfrac{V_0 + a(1-\tau)}{V_0}, & \frac{1}{2} < \tau \leqslant 1. \end{cases} \end{aligned}$$

Figure 25 shows the graph of $\tau \mapsto H(\tau)$ as well as the graph of the accumulation function for this cycle. Thus, on the time interval $(0,1)$ the net gain of heat is not zero; nevertheless, the relation (3.3) and the fact that $\mathbb{P}$ here is an isothermal path consisting of one segment and its reversal show that the accumulation function vanishes identically.

The distinction between distribution of heat with respect to time and with respect to temperature is crucial for the success of the present development. By emphasizing the distribution with respect to temperature or hotness as expressed through the accumulation function, we avoid the necessity of

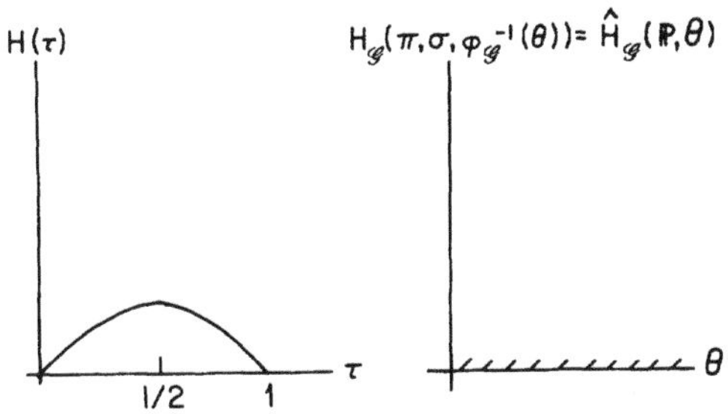

Figure 25.

giving details about how a system evolves in time. Whatever those details may be, it is only their ultimate contribution in determining an accumulation function which interests us. For example, the discussion in the previous paragraph leads us to conclude that infinitely many processes of an ideal gas have accumulation functions equal to the zero function.

We return now to the assertions (a), (b), and (c) about $H_{\mathscr{S}}$ and to the verbal forms 1–3 of the Second Law given in the introduction to this chapter. Each of these forms can now be interpreted as a statement of the following type: *for no cycle* (π, σ) *of* $\mathscr{S}$ *can the function* $L \mapsto H_{\mathscr{S}}(\pi, \sigma, L)$ *have a specified mathematical property*. For example, the Kelvin–Planck version 2 would read: for no cycle (π, σ) of $\mathscr{S}$ can $L \mapsto H_{\mathscr{S}}(\pi, \sigma, L)$ be increasing on some interval and nowhere be decreasing, i.e., the graph of $L \mapsto H_{\mathscr{S}}(\pi, \sigma, L)$ for a cycle (π, σ) cannot have the form shown in Figure 26. We leave it to the reader to supply similar statements of versions 1 and 3 in terms of the accumulation function for cycles of $\mathscr{S}$.

The concept of thermodynamical system thus yields simple, general, and precise statements of the Second Law which reflect the ideas contained in the classical verbal forms 1–3 of that law.

Although we are now in a position to compare these classical forms, we forego this in favor of giving a version of the Second Law due to Serrin. It turns out that Serrin's version not only implies the classical forms 1–3 but also can be analyzed in sufficient depth to provide a modern counterpart of

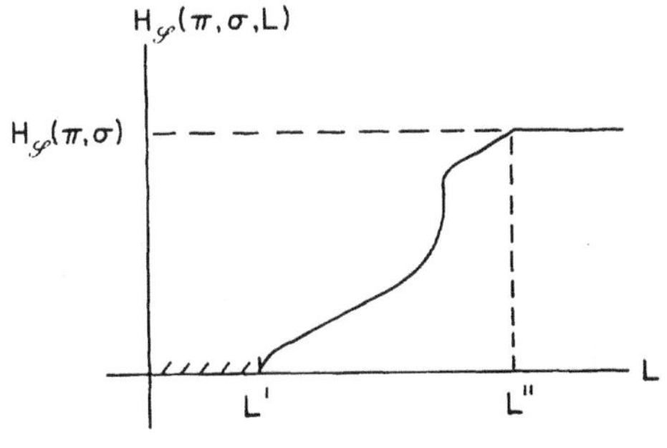

Figure 26.

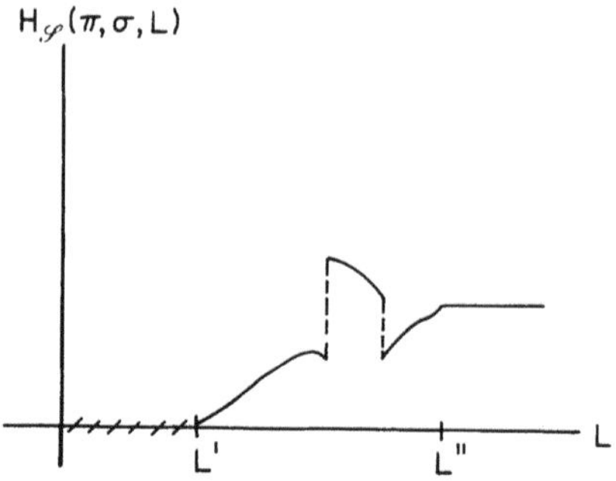

Figure 27.

the important "Clausius inequality" given in traditional treatments of the Second Law. This, in turn, will be the starting point for our study of the concept of entropy in Chapter V.

Second Law. *Let $\mathscr{S}$ be a thermodynamical system in the sense of Definition* 4.1. *For each cycle* (π, σ) *of* $\mathscr{S}$, *i.e., for each element of* $(\Pi_{\mathscr{S}} \Diamond \Sigma_{\mathscr{S}})_{\text{cyc}}$, *there holds*:

$$\text{if } H_{\mathscr{S}}(\pi, \sigma, \mathrm{L}) \geqslant 0 \quad \text{for all } \mathrm{L} \in \mathscr{M}, \text{ then } H_{\mathscr{S}}(\pi, \sigma) = 0. \tag{4.2}$$

According to (4.2), the accumulation function for a cycle of a system which obeys the Second Law cannot have the form shown in Figure 27, but may have the form shown in Figure 28. Actually, the main result on the Second Law, the Accumulation Theorem, implies (see Problem 6) that the only non-negative accumulation function for a cycle of a system obeying the Second Law is the zero function. In other words, the Second Law will imply that (4.2) can be replaced by:

$$\text{if } H_{\mathscr{S}}(\pi, \sigma, \mathrm{L}) \geqslant 0 \quad \text{for all } \mathrm{L} \in \mathscr{M}, \text{ then } H_{\mathscr{S}}(\pi, \sigma, \mathrm{L}) = 0 \quad \text{for all } \mathrm{L} \in \mathscr{M}. \tag{4.3}$$

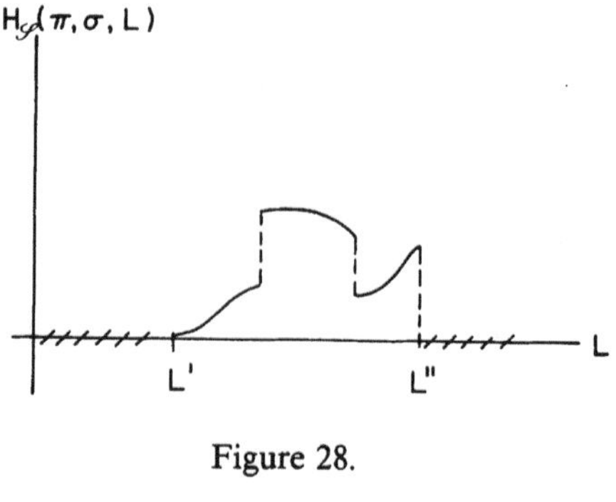

Figure 28.

5. Products of Systems and Preservation of the Second Law

The notion of a product of two systems was introduced in Definition 2.1 of Chapter III, and much of the presentation in Section 2 of Chapter III applies directly to our study of the Second Law. In fact, the only modifications of that presentation which are required here are ones which reflect the difference between the concept of thermodynamical system used in Chapter III and the one used in this chapter.

Definition 5.1. Let α be an index with values 1 and 2 and let $(\Sigma_{\mathscr{S}_\alpha}, \Pi_{\mathscr{S}_\alpha})$ and $H_{\mathscr{S}_\alpha}$ be the systems with perfect accessibility and accumulation functions specified for two thermodynamical systems $\mathscr{S}_1$ and $\mathscr{S}_2$. The *product* $\mathscr{S}_1 \times \mathscr{S}_2$ of $\mathscr{S}_1$ and $\mathscr{S}_2$ is defined through the following specification:

$$\Sigma_{\mathscr{S}_1 \times \mathscr{S}_2} = \Sigma_{\mathscr{S}_1} \times \Sigma_{\mathscr{S}_2}, \tag{5.1}$$

$$\Pi_{\mathscr{S}_1 \times \mathscr{S}_2} = \Pi_{\mathscr{S}_1} \times \Pi_{\mathscr{S}_2}, \tag{5.2}$$

and

$$H_{\mathscr{S}_1 \times \mathscr{S}_2}((\pi_1, \pi_2), (\sigma_1, \sigma_2), \mathrm{L}) = H_{\mathscr{S}_1}(\pi_1, \sigma_1, \mathrm{L}) + H_{\mathscr{S}_2}(\pi_2, \sigma_2, \mathrm{L}), \tag{5.3}$$

where (5.3) holds for every process (π_1, σ_1) of $\mathscr{S}_1$, (π_2, σ_2) of $\mathscr{S}_2$, and hotness level L.

The reader will notice that this definition differs from Definition 2.1 of Chapter III only in the replacement of (2.3) and (2.4) in Definition 2.1 by (5.3) in the present one. This difference arises because we here do not require the concept of work, but we do require more information about the distribution of heat gained with respect to hotness. Actually, because we have agreed to identify $H_{\mathscr{S}}(\pi, \sigma)$ with $H_{\mathscr{S}}(\pi, \sigma, \mathrm{L}'')$, relation (2.4) in Chapter III may be regarded as a consequence of (5.3) when L is chosen so that

$$L_1'' \preccurlyeq \mathrm{L} \quad \text{and} \quad \mathrm{L}_2'' \preccurlyeq \mathrm{L},$$

with L_1'' and L_2'' the upper hotness levels associated with (π_1, σ_1) and (π_2, σ_2), respectively.

Just as in Chapter III, we interpret $\mathscr{S}_1 \times \mathscr{S}_2$ as the thermodynamical system obtained by isolating $\mathscr{S}_1$ and $\mathscr{S}_2$ from one another and permitting them to operate simultaneously or sequentially without either thermal or mechanical interaction. The discussion in Chapter III can be easily adapted to this presentation in order to justify our calling $\mathscr{S}_1 \times \mathscr{S}_2$ a thermodynamical system in the sense of Definition 4.1.

Because every ideal gas $\mathscr{G}$ determines a thermodynamical system, it is meaningful to consider for each thermodynamical system $\mathscr{S}$ the product

$\mathscr{S} \times \mathscr{G}$. In complete analogy with our discussion in Section 2 of Chapter III, we here say that *$\mathscr{S}$ and $\mathscr{G}$ preserve the Second Law with respect to the product operation* if they satisfy the following condition:

$$\text{if } \mathscr{S} \text{ obeys the Second Law, then so does } \mathscr{S} \times \mathscr{G}. \tag{5.4}$$

6. The Second Law and the Accumulation Inequality

In proving property 3 of ideal gases in Section 3, we verified that every ideal gas satisfies (3.6), and we conclude from (3.3) that *every ideal gas $\mathscr{G}$ satisfies*

$$\int_0^{\infty} H_{\mathscr{G}}(\mathbb{P}, \varphi_{\mathscr{G}}^{-1}(\theta))\theta^{-2}\,d\theta = 0 \tag{6.1}$$

for every cycle $\mathbb{P}$ of $\mathscr{G}$. Our goal in this section is to study the relationship between the Second Law and a generalization of (6.1) to arbitrary thermodynamical systems. To state this generalization, we let $\mathscr{G}$ be an ideal gas and $\mathscr{S}$ be a thermodynamical system. We say that *$\mathscr{S}$ satisfies the Accumulation Inequality (for the temperature scale $\varphi_{\mathscr{G}}$ associated with $\mathscr{G}$)* if

$$(\pi, \sigma) \in (\Pi_{\mathscr{S}} \diamondsuit \Sigma_{\mathscr{S}})_{\text{cyc}} \Rightarrow \int_0^{\infty} H_{\mathscr{S}}(\pi, \sigma, \varphi_{\mathscr{G}}^{-1}(\theta))\theta^{-2}\,d\theta \leqslant 0. \tag{6.2}$$

The Accumulation Inequality plays a central role in the study of the Second Law, because it turns out to express the content of that law in a way which gives information about a system undergoing *arbitrary* cycles, whereas the Second Law mentions only the cycles for which the accumulation function is non-negative. Our presentation and study of the Accumulation Inequality is easily compared with our study of Joule's relation in Section 3 of Chapter III, if we make the following substitutions in the latter:

First Law → Second Law

Joule's relation → Accumulation Inequality.

We begin with an analogue of Lemma 3.1 of Chapter III.

Lemma 6.1. *If $\mathscr{S}$ satisfies the Accumulation Inequality for some choice of ideal gas $\mathscr{G}$, then $\mathscr{S}$ obeys the Second Law.*

PROOF. Suppose $\mathscr{S}$ satisfies (6.2) and let (π, σ) be a cycle of $\mathscr{S}$ such that, for all $\mathrm{L} \in \mathscr{M}$,

$$H_{\mathscr{S}}(\pi, \sigma, \mathrm{L}) \geqslant 0. \tag{6.3}$$

Let L″ be as in (TS1) for (π, σ), let $\theta'' = \varphi_{\mathscr{G}}(\mathrm{L}'')$ and note from (6.2) and

(6.3) that

$$\begin{aligned}0 &\geqslant \int_0^{\infty} H_{\mathscr{S}}\big(\pi,\sigma,\varphi_{\mathscr{G}}^{-1}(\theta)\big)\theta^{-2}\,d\theta\\ &= \int_0^{\theta''} H_{\mathscr{S}}\big(\pi,\sigma,\varphi_{\mathscr{G}}^{-1}(\theta)\big)\theta^{-2}\,d\theta + \int_{\theta''}^{\infty} H_{\mathscr{S}}\big(\pi,\sigma,\varphi_{\mathscr{G}}^{-1}(\theta)\big)\theta^{-2}\,d\theta\\ &\geqslant 0 + H_{\mathscr{S}}(\pi,\sigma)\int_{\theta''}^{\infty}\theta^{-2}\,d\theta\\ &= H_{\mathscr{S}}(\pi,\sigma)/\theta'' \geqslant 0.\end{aligned}$$

Therefore, we conclude that $\mathrm{H}_{\mathscr{S}}(\pi,\sigma)$ vanishes and that (4.2) holds, i.e., $\mathscr{S}$ obeys the Second Law. □

In the case where $\mathscr{S}$ is an ideal gas $\mathscr{G}$, relations (3.3) and (3.6) tell us that $\mathscr{S}$ does satisfy the Accumulation Inequality for the temperature scale $\varphi_{\mathscr{G}}$ associated with $\mathscr{G}$, and Lemma 6.1 implies that *every ideal gas obeys the Second Law.*

The next result is an analogue of Lemma 3.2 of Chapter III.

Lemma 6.2. *Let $\mathscr{S}$ be a thermodynamical system and $\mathscr{G}$ be an ideal gas. If $\mathscr{S}\times\mathscr{G}$ obeys the Second Law, then $\mathscr{S}$ satisfies the Accumulation Inequality for the temperature scale $\varphi_{\mathscr{G}}$ associated with $\mathscr{G}$.*

Proof. We shall show that if $\mathscr{S}$ *fails* to satisfy the Accumulation Inequality, then $\mathscr{S}\times\mathscr{G}$ does *not* obey the Second Law. Thus, we suppose that there is a cycle (π,σ) of $\mathscr{S}$ such that

$$\mathrm{I}_{\mathscr{S}}(\pi,\sigma):=\int_0^{\infty} H_{\mathscr{S}}\big(\pi,\sigma,\varphi_{\mathscr{G}}^{-1}(\theta)\big)\theta^{-2}\,d\theta > 0. \tag{6.4}$$

We wish to find a cycle $((\pi_{\mathscr{S}},\pi_{\mathscr{G}}),(\sigma_{\mathscr{S}},\sigma_{\mathscr{G}}))$ of $\mathscr{S}\times\mathscr{G}$ for which the implication (4.2) is false, i.e.,

$$H_{\mathscr{S}\times\mathscr{G}}\big((\pi_{\mathscr{S}},\pi_{\mathscr{G}}),(\sigma_{\mathscr{S}},\sigma_{\mathscr{G}}),\mathrm{L}\big)\geqslant 0 \quad \text{for all } \mathrm{L}\in\mathscr{M} \tag{6.5}$$

and

$$H_{\mathscr{S}\times\mathscr{G}}\big((\pi_{\mathscr{S}},\pi_{\mathscr{G}}),(\sigma_{\mathscr{S}},\sigma_{\mathscr{G}})\big) > 0. \tag{6.6}$$

Because every cycle $((\pi_{\mathscr{S}},\pi_{\mathscr{G}}),(\sigma_{\mathscr{S}},\sigma_{\mathscr{G}}))$ of $\mathscr{S}\times\mathscr{G}$ comes from cycles of $\mathscr{S}$ and cycles of $\mathscr{G}$, it is natural to take $\pi_{\mathscr{S}}=\pi$ and $\sigma_{\mathscr{S}}=\sigma$, with (π,σ) the cycle of $\mathscr{S}$ in (6.4). With this choice, (5.3) permits us to write (6.5) and (6.6) in the forms:

$$H_{\mathscr{S}}(\pi,\sigma,\mathrm{L})+H_{\mathscr{G}}(\pi_{\mathscr{G}},\sigma_{\mathscr{G}},\mathrm{L})\geqslant 0 \quad \text{for all } \mathrm{L}\in\mathscr{M} \tag{6.7}$$

and

$$H_{\mathscr{S}}(\pi,\sigma)+H_{\mathscr{G}}(\pi_{\mathscr{G}},\sigma_{\mathscr{G}}) > 0. \tag{6.8}$$

Now, let L′ and L″ be the upper and lower hotness levels in (TS1) for the cycle (π,σ), let $\theta_a=\varphi_{\mathscr{G}}(\mathrm{L}')$, $\theta_b=\varphi_{\mathscr{G}}(\mathrm{L}'')$, and assume that $(\pi_{\mathscr{G}},\sigma_{\mathscr{G}})$ has been chosen so that the path $\mathbb{P}$ corresponding to it lies on or between the lines $\theta = \theta_a$ and $\theta = \theta_b$. The crucial step is to express $H_{\mathscr{S}}(\pi, \sigma)$ and $H_{\mathscr{G}}(\pi_{\mathscr{G}}, \sigma_{\mathscr{G}})$ in (6.8) in terms of $\mathrm{I}_{\mathscr{S}}(\pi, \sigma)$ and

$$\mathrm{I}_{\mathscr{G}}(\mathbb{P}) := \int_0^{\infty} H_{\mathscr{G}}\big(\pi_{\mathscr{G}}, \sigma_{\mathscr{G}}, \varphi_{\mathscr{G}}^{-1}(\theta)\big)\theta^{-2}\, d\theta. \tag{6.9}$$

This can be done using the relations

$$\mathrm{I}_{\mathscr{S}}(\pi, \sigma) = H_{\mathscr{S}}(\pi, \sigma)\theta_b^{-1} + \int_{\theta_a}^{\theta_b} H_{\mathscr{S}}\big(\pi, \sigma, \varphi_{\mathscr{G}}^{-1}(\theta)\big)\theta^{-2}\, d\theta \tag{6.10}$$

and

$$\mathrm{I}_{\mathscr{G}}(\mathbb{P}) = H_{\mathscr{G}}(\pi_{\mathscr{G}}, \sigma_{\mathscr{G}})\theta_b^{-1} + \int_{\theta_a}^{\theta_b} H_{\mathscr{G}}\big(\pi_{\mathscr{G}}, \sigma_{\mathscr{G}}, \varphi_{\mathscr{G}}^{-1}(\theta)\big)\theta^{-2}\, d\theta, \tag{6.11}$$

so that (6.10), (6.11), (3.3) and (3.6) yield

$$\begin{aligned} H_{\mathscr{S}}(\pi, \sigma) + H_{\mathscr{G}}(\pi_{\mathscr{G}}, \sigma_{\mathscr{G}}) &= \theta_b \mathrm{I}_{\mathscr{S}}(\pi, \sigma) - \theta_b \int_{\theta_a}^{\theta_b} H_{\mathscr{S}}\big(\pi, \sigma, \varphi_{\mathscr{G}}^{-1}(\theta)\big)\theta^{-2}\, d\theta \\ &\quad + \theta_b \cdot 0 - \theta_b \int_{\theta_a}^{\theta_b} H_{\mathscr{G}}\big(\pi_{\mathscr{G}}, \sigma_{\mathscr{G}}, \varphi_{\mathscr{G}}^{-1}(\theta)\big)\theta^{-2}\, d\theta \\ &= \theta_b \mathrm{I}_{\mathscr{S}}(\pi, \sigma) - \theta_b \int_{\theta_a}^{\theta_b} \theta^{-2} A(\theta)\, d\theta, \end{aligned} \tag{6.12}$$

where

$$A(\theta) = H_{\mathscr{S}}\big(\pi, \sigma, \varphi_{\mathscr{G}}^{-1}(\theta)\big) + H_{\mathscr{G}}\big(\pi_{\mathscr{G}}, \sigma_{\mathscr{G}}, \varphi_{\mathscr{G}}^{-1}(\theta)\big). \tag{6.13}$$

Because the right hand-side of (6.13) is just another way of writing the left-hand side of the inequality in (6.7), we see from (6.12) and (6.13) that (6.7) and (6.8) will hold if we can find $(\pi_{\mathscr{G}}, \sigma_{\mathscr{G}})$, a cycle of $\mathscr{G}$, such that

$$A(\theta) \geqslant 0 \quad \text{for all } \theta > 0 \tag{6.14}$$

and

$$\frac{A(\theta_b)}{\theta_b} = \mathrm{I}_{\mathscr{S}}(\pi, \sigma) - \int_{\theta_a}^{\theta_b} \theta^{-2} A(\theta)\, d\theta > 0. \tag{6.15}$$

However, if $A(\theta) \geq 0$ holds for θ in (θ_a, θ_b), then

$$\int_{\theta_a}^{\theta_b} \theta_a^{-2} A(\theta)\, d\theta \geqslant \int_{\theta_a}^{\theta_b} \theta^{-2} A(\theta)\, d\theta,$$

and (6.15) will follow from the inequality

$$\mathrm{I}_{\mathscr{S}}(\pi, \sigma) - \theta_a^{-2} \int_{\theta_a}^{\theta_b} A(\theta)\, d\theta > 0.$$

In other words, (6.7) and (6.8) are implied by the relations (6.14) and

$$\theta_a^2 \mathrm{I}_{\mathscr{S}}(\pi, \sigma) > \int_{\theta_a}^{\theta_b} A(\theta)\, d\theta. \tag{6.16}$$

The inequalities (6.14) and (6.16) tell us that the cycle $(\pi_{\mathscr{G}}, \sigma_{\mathscr{G}})$ should be chosen so that the function A is non-negative on all of $\mathbb{R}^{++}$ with its integral from θ_a to θ_b smaller than the positive number $\theta_a^2 \mathrm{I}_{\mathscr{S}}(\pi, \sigma)$.

To focus attention on the part of A which is determined by $(\pi_{\mathscr{G}}, \sigma_{\mathscr{G}})$ alone, we restate our conclusions as follows: $(\pi_{\mathscr{G}}, \sigma_{\mathscr{G}})$ must be chosen so that

ⓐ the closed path $\mathbb{P}$ which is determined by $(\pi_{\mathscr{G}}, \sigma_{\mathscr{G}})$ is bounded by the lines $\theta = \theta_a$ and $\theta = \theta_b$;

ⓑ $\hat{H}_{\mathscr{G}}(\mathbb{P}, \theta) \geqslant f(\theta), \qquad 0 < \theta < \infty$;

ⓒ $\varepsilon_0 + \int_{\theta_a}^{\theta_b} f(\theta)\, d\theta > \int_{\theta_a}^{\theta_b} \hat{H}_{\mathscr{G}}(\mathbb{P}, \theta)\, d\theta.$

In ⓑ and ⓒ we have written $f(\theta)$ for $-H_{\mathscr{S}}(\pi, \sigma, \varphi_{\mathscr{G}}^{-1}(\theta))$ and ε_0 for the positive number $\theta_a^2 \mathrm{I}_{\mathscr{S}}(\pi, \sigma)$; we also have used (3.3) to replace $H_{\mathscr{G}}(\pi_{\mathscr{G}}, \sigma_{\mathscr{G}}, \varphi_{\mathscr{G}}^{-1}(\theta))$ by $\hat{H}_{\mathscr{G}}(\mathbb{P}, \theta)$. In order to proceed further, we observe that $-f$ is the accumulation function for (π, σ) expressed as a function of θ, so we may apply (TS2) in Definition 4.1 to conclude that the restriction of f to $[\theta_a, \theta_b)$ is integrable in the sense of Riemann. The theory of the Riemann integral tells us that there is a step function q on $[\theta_a, \theta_b)$ such that

$$q(\theta) \geqslant f(\theta), \qquad \theta_a \leqslant \theta < \theta_b \tag{6.17}$$

and

$$\varepsilon_0 + \int_{\theta_a}^{\theta_b} f(\theta)\, d\theta > \int_{\theta_a}^{\theta_b} q(\theta)\, d\theta. \tag{6.18}$$

Clearly, (6.17) and (6.18) would yield ⓒ and ⓑ (for $\theta_a \leqslant \theta < \theta_b$ instead of $0 < \theta < \infty$) if we could replace $q(\theta)$ by $\hat{H}_{\mathscr{G}}(\mathbb{P}, \theta)$ for some choice of cycle $\mathbb{P}$. To this end, we observe that q can be modified so that (6.17) and (6.18) still hold *and* q is right continuous, and we are now able to apply the Lemma of Section 3 to obtain the existence of a cycle $\mathbb{P}$ of $\mathscr{G}$ satisfying ⓐ, ⓒ, and ⓑ—for $\theta_a \leqslant \theta < \theta_b$. However, these conclusions together imply that ⓑ also holds for $\theta_b \leqslant \theta$, because $H_{\mathscr{S}}(\pi, \sigma, \varphi_{\mathscr{G}}^{-1}(\theta))$ and $\hat{H}_{\mathscr{G}}(\mathbb{P}, \theta)$ are constant on $[\theta_b, \infty)$ and because, in view of these conclusions, ⓒ in the form (6.16) now implies (6.15). Moreover, both $\hat{H}_{\mathscr{G}}$ and f vanish on $(0, \theta_a)$, so that ⓑ is verified. Hence, ⓐ, ⓑ, and ⓒ hold, which implies that so do (6.14) and (6.15); consequently, (6.5) and (6.6) hold for the cycle $((\pi, \pi_{\mathscr{G}}), (\sigma, \sigma_{\mathscr{G}}))$ just constructed, and this completes the proof of Lemma 6.2. □

Lemmas 6.1 and 6.2 yield the following pair of implications:

$\mathscr{S} \times \mathscr{G}$ obeys the Second Law $\Rightarrow$ $\mathscr{S}$ satisfies the Accumulation Inequality

$\Rightarrow$ $\mathscr{S}$ obeys the Second Law.

The situation here is the same as in our analysis of the First Law in Chapter III; we continue here in the same way as in Chapter III and state the next result as a theorem.

Theorem 6.1 (Accumulation Theorem). *Let $\mathscr{S}$ be a thermodynamical system, let $\mathscr{G}$ be an ideal gas, and suppose that $\mathscr{S}$ and $\mathscr{G}$ preserve the Second Law with respect to the product operation. It follows that $\mathscr{S}$ obeys the Second Law if and only if $\mathscr{S}$ satisfies the Accumulation Inequality (for the temperature scale $\varphi_{\mathscr{G}}$).*

PROOF. If $\mathscr{S}$ obeys the Second Law, then by (5.4) so does $\mathscr{S}\times\mathscr{G}$, and Lemma 6.2 tells us that $\mathscr{S}$ satisfies the Accumulation Inequality. Conversely, if $\mathscr{S}$ satisfies the Accumulation Inequality, then $\mathscr{S}$ obeys the Second Law, by Lemma 6.1. □

The Accumulation Theorem has many important consequences, some of which we discuss here and in the next section. First of all, we may ask whether or not the hypothesis of the theorem is satisfied when $\mathscr{S}$ is an ideal gas, $\mathscr{S}=\mathscr{G}^*$, possibly different from $\mathscr{G}$. If $\mathscr{G}^*$ and $\mathscr{G}$ preserve the Second Law with respect to the product operation, then, because every ideal gas does obey the Second Law, the Accumulation Theorem tells us that $\mathscr{G}^*$ satisfies the Accumulation Inequality for the temperature scale $\varphi_{\mathscr{G}}$ associated with $\mathscr{G}$:

$$(\pi_{\mathscr{G}^*},\sigma_{\mathscr{G}^*})\in(\Pi_{\mathscr{G}^*}\diamondsuit\Sigma_{\mathscr{G}^*})_{\mathrm{cyc}}\Rightarrow\int_0^\infty H_{\mathscr{G}^*}\big(\pi_{\mathscr{G}^*},\sigma_{\mathscr{G}^*},\varphi_{\mathscr{G}}^{-1}(\theta)\big)\theta^{-2}\,d\theta\leqslant 0. \tag{6.19}$$

However, $H_{\mathscr{G}^*}(\pi_{\mathscr{G}^*},\sigma_{\mathscr{G}^*},\mathrm{L})$ for every hotness level L is given by the formula

$$H_{\mathscr{G}^*}(\pi_{\mathscr{G}^*},\sigma_{\mathscr{G}^*},\mathrm{L})=\int_{\{\tau|\bar\theta(\tau)\leqslant\varphi_{\mathscr{G}^*}(\mathrm{L})\}}\bar{\mathfrak{h}}(\tau)\,d\tau, \tag{6.20}$$

where $\bar{\mathfrak{h}}$ is the heating for a standard parameterization $(\bar V,\bar\theta)$ of the path $\mathbb{P}^*$ determined by the cycle $(\pi_{\mathscr{G}^*},\sigma_{\mathscr{G}^*})$. By employing (6.20) with L replaced by $\varphi_{\mathscr{G}}^{-1}(\theta)$ and using the notation

$$T(\theta)=\varphi_{\mathscr{G}}\big(\varphi_{\mathscr{G}^*}^{-1}(\theta)\big),\qquad 0<\theta<\infty, \tag{6.21}$$

we conclude that (6.19) is equivalent to the assertion

$$(\pi_{\mathscr{G}^*},\sigma_{\mathscr{G}^*})\in(\Pi_{\mathscr{G}^*}\diamondsuit\Sigma_{\mathscr{G}^*})_{\mathrm{cyc}}\Rightarrow\int_0^\infty\left(\int_{\{\tau|T(\bar\theta(\tau))\leqslant\theta\}}\bar{\mathfrak{h}}(\tau)\,d\tau\right)\theta^{-2}\,d\theta\leqslant 0. \tag{6.22}$$

The argument used to verify (3.5) yields the relation

$$\int_0^\infty\int_{\{\tau|T(\bar\theta(\tau))\leqslant\theta\}}\bar{\mathfrak{h}}(\tau)\theta^{-2}\,d\tau\,d\theta=\int_0^1\frac{\bar{\mathfrak{h}}(\tau)}{T(\bar\theta(\tau))}\,d\tau. \tag{6.23}$$

Moreover, the formula

$$\bar{\mathfrak{h}}(\tau)=\lambda(\bar V(\tau),\bar\theta(\tau))\bar V^{\cdot}(\tau)+\varkappa(\bar V(\tau),\bar\theta(\tau))\bar\theta^{\cdot}(\tau)$$

in the case of an ideal gas $\mathscr{G}^*$ reduces to

$$\bar{\mathfrak{h}}(\tau)=\lambda^*\bar\theta(\tau)\frac{\bar V^{\cdot}(\tau)}{\bar V(\tau)}+\varkappa^*(\bar\theta(\tau))\bar\theta^{\cdot}(\tau),$$

so that (6.22) is equivalent to the statement that

$$\oint_{\mathbb{P}^*} \frac{\lambda^* \theta}{T(\theta) V} dV + \frac{\partial^*(\theta)}{T(\theta)} d\theta \leqslant 0 \tag{6.24}$$

for every cycle $\mathbb{P}^*$ of $\mathscr{G}^*$. In particular, if we apply (6.24) to $\mathbb{P}_r^*$, the reversal of a cycle $\mathbb{P}^*$ of $\mathscr{G}^*$, then the fact that the line integral in (6.24) changes sign when $\mathbb{P}^*$ is replaced by $\mathbb{P}_r^*$ tells us that (6.22) is equivalent to the condition

$$\int_{\mathbb{P}^*} \frac{\lambda^* \theta}{T(\theta) V} dV + \frac{\partial^*(\theta)}{T(\theta)} d\theta = 0 \tag{6.25}$$

for all cycles $\mathbb{P}^*$ of $\mathscr{G}^*$. If we apply (6.25) to a rectangular path $\mathbb{P}^*$ with sides parallel to the coordinate axes, we conclude immediately that $\theta \mapsto \theta / T(\theta)$ is a constant function. Consequently, (6.21) becomes

$$c\theta = \varphi_{\mathscr{G}}\big(\varphi_{\mathscr{G}^*}^{-1}(\theta)\big), \qquad 0 < \theta < \infty \tag{6.26}$$

where c is a fixed positive number, and this relation with θ replaced by $\varphi_{\mathscr{G}^*}(\mathrm{L})$ becomes

$$c\varphi_{\mathscr{G}^*}(\mathrm{L}) = \varphi_{\mathscr{G}}(\mathrm{L}), \qquad \mathrm{L} \in \mathscr{M}. \tag{6.27}$$

In summary, if $\mathscr{G}^*$ and $\mathscr{G}$ preserve the Second Law with respect to the product operation, then the corresponding empirical temperature scales $\varphi_{\mathscr{G}^*}$ and $\varphi_{\mathscr{G}}$ are positive constant multiples of one another. Conversely, if (6.27) or, equivalently, (6.26) holds, then we have for each cycle $\mathbb{P}^*$ of $\mathscr{G}^*$ and $\mathbb{P}$ of $\mathscr{G}$:

$$\begin{aligned}
&\int_0^\infty \frac{H_{\mathscr{G}^*}\big(\mathbb{P}^*, \varphi_{\mathscr{G}}^{-1}(\theta)\big) + H_{\mathscr{G}}\big(\mathbb{P}, \varphi_{\mathscr{G}}^{-1}(\theta)\big)}{\theta^2} d\theta \\
&\quad = \int_0^\infty \frac{H_{\mathscr{G}^*}\big(\mathbb{P}^*, \varphi_{\mathscr{G}^*}^{-1}(c^{-1}\theta)\big)}{\theta^2} d\theta + \int_0^\infty \frac{H_{\mathscr{G}}\big(\mathbb{P}, \varphi_{\mathscr{G}}^{-1}(\theta)\big)}{\theta^2} d\theta \\
&\quad = c^{-1} \int_0^\infty \frac{H_{\mathscr{G}^*}\big(\mathbb{P}^*, \varphi_{\mathscr{G}^*}^{-1}(\theta^*)\big)}{(\theta^*)^2} d\theta^* + \int_0^\infty \frac{H_{\mathscr{G}}\big(\mathbb{P}, \varphi_{\mathscr{G}}^{-1}(\theta)}{\theta^2} d\theta \\
&\quad = 0,
\end{aligned}$$

i.e., the product $\mathscr{G}^* \times \mathscr{G}$ satisfies the Accumulation Inequality for $\varphi_{\mathscr{G}}$. By Lemma 6.1, $\mathscr{G}^* \times \mathscr{G}$ obeys the Second Law, and this shows that $\mathscr{G}^*$ and $\mathscr{G}$ preserve the Second Law with respect to the product operation. Thus, we have proven that *two ideal gases preserve the Second Law with respect to the product operation if and only if their temperature scales are constant multiples of one another.*

For later use we note that the reasoning which led from (6.22) to (6.27) yields the following result.

Lemma 6.3. *Let $\mathscr{G}$ be an ideal gas, and let φ be an empirical temperature scale with range equal to $\mathbb{R}^{++}$. It follows that $\mathscr{G}$ satisfies the Accumulation Inequal-*

ity for the temperature scale φ [*i.e.*, (6.2) *holds with* $\mathscr{S}$ *replaced by* $\mathscr{G}$ *and* $\varphi_{\mathscr{G}}$ *replaced by* φ] *if and only if* φ *is a constant multiple of* $\varphi_{\mathscr{G}}$.

One of the principal uses of the Accumulation Inequality concerns finding the consequences of the Second Law for various collections of thermodynamical systems. This is best done when the same temperature scale occurs in the Accumulation Inequality for all of the systems under consideration. The next result gives conditions which imply that this is so.

Theorem 6.2. *Let* $\mathbb{U}$ *be a collection of thermodynamical systems, and let* $\mathscr{G}$ *be an ideal gas with associated temperature scale* $\varphi_{\mathscr{G}}$. *Suppose that not only every system* $\mathscr{S}$ *in* $\mathbb{U}$ *but also all the products* $\mathscr{S} \times \mathscr{G}$ *with* $\mathscr{S}$ *in* $\mathbb{U}$ *obey the Second Law. It follows that every system* $\mathbb{U}$ *satisfies the Accumulation Inequality for the scale* $\varphi_{\mathscr{G}}$.

PROOF. The fact that the systems $\mathscr{S}$ *and* $\mathscr{S} \times \mathscr{G}$ obey the Second Law implies that, for each $\mathscr{S}$ in $\mathbb{U}$, the systems $\mathscr{S}$ and $\mathscr{G}$ preserve the Second Law with respect to the product operation. The conclusion of the theorem now follows from the Accumulation Theorem. □

The condition that all products of the form $\mathscr{S} \times \mathscr{G}$ obey the Second Law is the counterpart of Carnot's condition that two heat engines operating together should obey a primitive version of that law. Thus, both in classical and modern treatments, the usefulness of the Second Law depends upon its validity for pairs of systems when regarded as single systems.

The next result together with the previous one lead us to grant a special status to the temperature scale $\varphi_{\mathscr{G}}$ associated with an ideal gas $\mathscr{G}$. This is usually done in traditional presentations by calling $\varphi_{\mathscr{G}}$, or any of its constant multiples, an "absolute temperature scale."

Theorem 6.3. *Let* $\mathbb{U}$ *be a collection of thermodynamical systems which contains an ideal gas* $\mathscr{G}$, *and let* φ *be an empirical temperature scale with range equal to* $\mathbb{R}^{++}$. *If every system in* $\mathbb{U}$ *satisfies the Accumulation Inequality for the scale* φ, *then* φ *is a constant multiple of* $\varphi_{\mathscr{G}}$, *the temperature scale associated with* $\mathscr{G}$.

PROOF. By hypothesis, $\mathscr{G}$ satisfies the Accumulation Inequality for the scale φ, and Lemma 6.3 tells us that φ is then a constant multiple of $\varphi_{\mathscr{G}}$. □

In suggestive terms, Theorems 6.2 and 6.3 tell us that a collection of systems which is "compatible" with the Second Law and which contains at least one ideal gas $\mathscr{G}$ satisfies the Accumulation Inequality for one and (to within constant multiples) only one temperature scale, namely $\varphi_{\mathscr{G}}$. It is this remarkable property of $\varphi_{\mathscr{G}}$ which leads us to call it an *absolute temperature scale*.

In the discussion of temperature given in Chapter I just after Definition 1.1, we mentioned that the choice of temperature scale made there, and employed throughout that chapter, was subject to a restriction to be explained here in Chapter IV. It is clear now that the temperature scale used in Chapter I must be an absolute temperature scale. In fact, Theorem 6.3 can be applied to the class $\mathbb{U}_c$ of homogeneous fluid bodies which satisfy the classical versions of the First and Second Laws, because Theorems 1.2 and 3.1 of Chapter I imply that every system in $\mathbb{U}_c$ satisfies the Accumulation Inequality for the chosen temperature scale, and $\mathbb{U}_c$ contains all ideal gases.

It is interesting to note that, had we not employed an ideal gas in our analysis of the Second Law, but instead had we used the Second Law to construct, as Serrin has done, a temperature scale φ which satisfies the hypothesis of Theorem 6.3, the scale φ would have turned out to be a multiple of $\varphi_{\mathscr{G}}$. It is only when one wishes to study collections of systems which do not contain at least one ideal gas that such an alternative approach to the analysis of the Second Law becomes significant.

7. Estimates for Heat Absorbed and Heat Emitted

In order to illustrate a further application of the Accumulation Inequality, we show here how estimates similar in form to those obtained in Theorem 4.1 of Chapter I for homogeneous fluid bodies can be derived easily for general thermodynamical systems from (6.2).

Our basic hypothesis will be that, for each process (π, σ) of a system $\mathscr{S}$, the function $\theta \mapsto H_{\mathscr{S}}(\pi, \sigma, \varphi_{\mathscr{G}}^{-1}(\theta))$ in (6.2) can be written as the difference of two non-negative and non-decreasing functions, i.e., for each θ in $\mathbb{R}^{++}$,

$$H_{\mathscr{S}}\big(\pi, \sigma, \varphi_{\mathscr{G}}^{-1}(\theta)\big) = H_{\mathscr{S}}^{\downarrow}(\pi, \sigma, \theta) - H_{\mathscr{S}}^{\uparrow}(\pi, \sigma, \theta). \qquad (7.1)$$

We assume also that $H_{\mathscr{S}}^{\downarrow}$ and $H_{\mathscr{S}}^{\uparrow}$ are zero on $(0, \varphi_{\mathscr{G}}(\mathrm{L}'))$ and are constant on $[\varphi_{\mathscr{G}}(\mathrm{L}''), \infty)$. Results in Analysis tell us that such decompositions always exist and single out particular ones in which the terms have special properties. Here, our concern is not how these decompositions arise but, rather, how they can be used. We call $H_{\mathscr{S}}^{\downarrow}(\pi, \sigma, \theta)$ the *net heat absorbed at or below temperature* θ in (π, σ) and $H_{\mathscr{S}}^{\uparrow}(\pi, \sigma, \theta)$ the *net heat emitted at or below* θ. These names are used to help us fix the physical interpretation of (7.1) but they will not be justified here. When (π, σ) is a cycle of $\mathscr{S}$, the Accumulation Inequality (6.2) yields

$$\begin{aligned} 0 &\geqslant \int_0^\infty H_{\mathscr{S}}\big(\pi, \sigma, \varphi_{\mathscr{G}}^{-1}(\theta)\big)\theta^{-2}\, d\theta \\ &= \int_0^\infty H_{\mathscr{S}}^{\downarrow}(\pi, \sigma, \theta))\theta^{-2}\, d\theta - \int_0^\infty H_{\mathscr{S}}^{\uparrow}(\pi, \sigma, \theta)\theta^{-2}\, d\theta \qquad (7.2) \\ &\geqslant \int_{\theta''}^\infty H_{\mathscr{S}}^{\downarrow}(\pi, \sigma, \theta)\theta^{-2}\, d\theta - \int_{\theta'}^\infty H_{\mathscr{S}}^{\uparrow}(\pi, \sigma, \theta)\theta^{-2}\, d\theta, \end{aligned}$$

where $\theta'' = \varphi_{\mathscr{G}}(\mathrm{L}'')$ and $\theta' = \varphi_{\mathscr{G}}(\mathrm{L}')$. In obtaining the inequality (7.2), we have used the fact that $H_{\mathscr{S}}^{\downarrow}$ and $H_{\mathscr{S}}^{\uparrow}$ both vanish on $(0, \theta')$ and $H_{\mathscr{S}}^{\downarrow}$ is non-negative. The last integral in (7.2) satisfies

$$-\int_{\theta'}^{\infty} H_{\mathscr{S}}^{\uparrow}(\pi, \sigma, \theta)\theta^{-2}\, d\theta \geqslant -\int_{\theta'}^{\infty} H_{\mathscr{S}}^{\uparrow}(\pi, \sigma, \theta'')\theta^{-2}\, d\theta$$

because $H_{\mathscr{S}}^{\uparrow}$ is non-decreasing, while the next-to-last satisfies

$$\int_{\theta''}^{\infty} H_{\mathscr{S}}^{\downarrow}(\pi, \sigma, \theta)\theta^{-2}\, d\theta = H_{\mathscr{S}}^{\downarrow}(\pi, \sigma, \theta'')\int_{\theta''}^{\infty} \theta^{-2}\, d\theta.$$

These relations and (7.2) immediately yield the estimate

$$0 \geqslant \frac{H_{\mathscr{S}}^{\downarrow}(\pi, \sigma, \theta'')}{\theta''} - \frac{H_{\mathscr{S}}^{\uparrow}(\pi, \sigma, \theta'')}{\theta'},$$

i.e., if $H_{\mathscr{S}}^{\uparrow}(\pi, \sigma, \theta'')$ is non-zero,

$$\frac{H_{\mathscr{S}}^{\downarrow}(\pi, \sigma, \theta'')}{H_{\mathscr{S}}^{\uparrow}(\pi, \sigma, \theta'')} \leqslant \frac{\theta''}{\theta'}. \tag{7.3}$$

This inequality is reminiscent of (4.7), Chapter I:

$$\frac{H^{+}(\mathbb{P})}{H^{-}(\mathbb{P})} \leqslant \frac{\theta_2}{\theta_1},$$

which holds for cycles $\mathbb{P}$ of a homogeneous fluid body during which no heat is absorbed above temperature θ_2 and no heat is emitted below θ_1. This inequality was shown to reduce to equality if and only if $\mathbb{P}$ is a Carnot path on which heat is emitted only at temperature θ_1 and absorbed only at θ_2. An analogous argument shows here that equality holds in (7.3) if and only if $H_{\mathscr{S}}(\pi, \sigma, \varphi_{\mathscr{G}}^{-1}(\theta))$ is given by

$$H_{\mathscr{S}}(\pi, \sigma, \varphi_{\mathscr{G}}^{-1}(\theta)) = \begin{cases} 0, & \theta < \theta' \\ A, & \theta' \leqslant \theta < \theta'' \\ B, & \theta'' \leqslant \theta, \end{cases} \tag{7.4}$$

with A and B real numbers related by the formula

$$B = A\left(1 - \frac{\theta''}{\theta'}\right). \tag{7.5}$$

The reader should compare (7.4) and (7.5) with the formula (3.4) for the accumulation function of an ideal gas for a Carnot cycle. That comparison strengthens the analogy between the estimate in classical thermodynamics ((4.7) of Chapter I) and the inequality (7.3), and it shows that little has been sacrificed to gain the greater generality of the modern treatment.

CHAPTER V

Energy and Entropy for Thermodynamical Systems

1. Introduction

In our treatment of homogeneous fluid bodies, we found that the laws of thermodynamics permitted us to augment the list of functions of state $(\mathscr{p}, \tilde{\lambda}, \mathscr{c})$ in the description of such a body. The new functions of state, E and S—called energy and entropy functions, respectively—arise as potentials for vector fields on the state space, their existence is equivalent to the First and Second Laws, and their derivatives yield the original functions of state, $\mathscr{p}$, $\tilde{\lambda}$, and $\mathscr{c}$ in simple relations. In Chapters III and IV we have studied versions of the First and Second Laws which are meaningful for a class of systems far broader than the collection of homogeneous fluid bodies and which take the form of relations that restrict the behavior of a thermodynamical system in special cycles. For each of the two laws of thermodynamics, we found a condition that applies to all cycles and is equivalent to the original statement of the law. Theorem 3.1 of Chapter III and Theorem 6.1 of Chapter IV express these facts in precise mathematical language and tell us that Joule's relation, (3.1) of Chapter III (with $M = \mathrm{R}/\lambda$), is equivalent to the First Law, and that the Accumulation Inequality, (6.2) of Chapter IV, is equivalent to the Second Law. It is natural to ask whether analogues of energy and entropy can be obtained from Joule's relation and the Accumulation Inequality, and the present chapter is devoted to providing an affirmative answer to this question.

Before embarking on the main part of the presentation, we pause to explain the point of view toward energy and entropy which will emerge here. Energy and entropy are *derived* quantities in the present theory, as distinct from heat, hotness and work, which may be described as *primitive* quantities, i.e., ones which are given at the outset and are not defined in terms of other concepts. In fact, we will show that the laws of thermodynamics imply the existence of functions of state which obey relations that are generalizations of (3.27) and (3.28) of Chapter I, and we will then name these functions "energy functions" and "entropy functions", because of the close analogy between the relations they satisfy and the relations (3.27) and (3.28). The importance of energy and entropy functions lies in the fact that they permit us to state the entire content of the First and Second Laws as conditions which hold for *every* process of a system, in contrast with the initial statements of these laws as conditions which hold only for special cyclic processes.

The relationship between the laws of thermodynamics for thermodynamical systems and the existence of energy and entropy functions rests only on part of the structure described in Definition 1.1 of Chapter III and in Definition 4.1 of Chapter IV, namely, the notion of a system with perfect accessibility. In this chapter we shall use the fact that both Joule's relation and the Accumulation Inequality are statements of the following form: a preassigned action for a system with perfect accessibility vanishes on every cycle or is not positive on every cycle of the system. The fact that the action in Joule's relation is of the form $(\pi,\sigma)\mapsto W_{\mathscr{S}}(\pi,\sigma)-(\mathrm{R}/\lambda)H_{\mathscr{S}}(\pi,\sigma)$ and that the action $(\pi,\sigma)\mapsto \mathrm{I}_{\mathscr{S}}(\pi,\sigma)$ in the Accumulation Inequality is an integral of the accumulation function of $\mathscr{S}$ is not used in our theory of energy and entropy functions. Thus, we may regard the analysis of Chapters III and IV as having singled out actions with special properties on the cycles of a system with perfect accessibility, and we here study in detail the class of actions having these properties.

2. Actions with the Clausius Property

In both classical and modern treatments of the Second Law the content of that law is equivalent to a statement that a given action for a system with perfect accessibility not be positive on cycles of the system. For the modern treatment this is clear from Theorem 6.1 of Chapter IV; for the classical treatment, in which the Second Law reduces to the assertion that (3.24), Chapter I, holds on all cycles, one must observe that the assertion

$$\mathbb{P} \text{ a cycle} \Rightarrow \oint_{\mathbb{P}} \frac{\tilde{\lambda}(V,\theta)}{\theta}dV + \frac{\mathscr{s}(V,\theta)}{\theta}d\theta \leqslant 0 \tag{2.1}$$

is equivalent to the statement

$$\mathbb{P} \text{ a cycle} \Rightarrow \oint_{\mathbb{P}} \frac{\tilde{\lambda}(V,\theta)}{\theta}dV + \frac{\mathscr{s}(V,\theta)}{\theta}d\theta = 0, \tag{2.2}$$

which is precisely (3.24) of Chapter I. The proof of the equivalence of (2.1) and (2.2) is left as an exercise; it is based on the fact that every cycle $\mathbb{P}$ of a homogeneous fluid body has a reversal $\mathbb{P}_r$ which also is a cycle.

All of the presentation in the remainder of this chapter rests on the concepts in Chapter II, and it may be useful at this point for the reader to review Definitions 1.1 and 2.1 of that chapter. We now describe the principal new concept of this section.

Definition 2.1. Let a be an action for a system (Σ, Π) with perfect accessibility, and let σ° be in Σ. We say that a has the *Clausius property at* σ° if there holds

$$(\pi, \sigma^\circ) \in (\Pi \diamond \Sigma)_{\text{cyc}} \Rightarrow a(\pi, \sigma^\circ) \leqslant 0. \tag{2.3}$$

Our choice of the term "Clausius property" rests on the classical term "Clausius inequality" for a special instance of (2.3). Coleman and Owen employed the term "Clausius property" to describe actions which satisfy an approximate version of (2.3) in the context of a theory of systems with "approximate cycles." The remainder of this section contains an adaptation of Coleman and Owen's study of actions with the Clausius property to the systems with perfect accessibility studied here. We present several preliminary results in a series of lemmas. In each of these lemmas, a is an action for a system with perfect accessibility (Σ, Π).

Lemma 2.1. *a has the Clausius property at a state σ° if and only if, for each σ in Σ, the set of real numbers*

$$a\{\sigma^\circ \to \sigma\} := \{a(\pi, \sigma^\circ) \mid \rho_\pi \sigma^\circ = \sigma\} \tag{2.4}$$

is bounded above.

PROOF. Suppose that a has the Clausius property at σ° and let σ be in Σ. Property (S1) in Definition 1.1, Chapter II, implies there is a process generator $\bar{\pi}$ such that

$$\rho_{\bar{\pi}} \sigma = \sigma^\circ \tag{2.5}$$

(see Figure 29). For each π such that $\rho_\pi \sigma^\circ = \sigma$, there holds

$$\sigma \in \mathscr{D}(\bar{\pi}) \cap \mathscr{R}(\pi),$$

so that $\mathscr{D}(\bar{\pi}) \cap \mathscr{R}(\pi) \neq \varnothing$ and $\bar{\pi}\pi$ is a process generator with $\sigma^\circ \in \mathscr{D}(\bar{\pi}\pi)$. Moreover, (1.3) of Chapter II, (2.5), and the choice of π imply that

$$\rho_{\bar{\pi}\pi} \sigma^\circ = \rho_{\bar{\pi}} \rho_\pi \sigma^\circ = \rho_{\bar{\pi}} \sigma = \sigma^\circ,$$

and we conclude that $(\bar{\pi}\pi, \sigma^\circ)$ is a cycle. Because a has the Clausius property at σ°, (2.3) here and (2.1) of Chapter II yield

$$\begin{aligned} 0 \geqslant a(\bar{\pi}\pi, \sigma^\circ) &= a(\pi, \sigma^\circ) + a(\bar{\pi}, \rho_\pi \sigma^\circ) \\ &= a(\pi, \sigma^\circ) + a(\bar{\pi}, \sigma). \end{aligned}$$

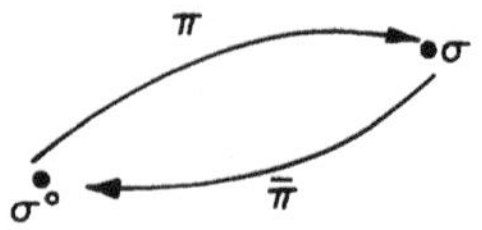

Figure 29.

Therefore, for every process generator π for which $\rho_\pi\sigma^\circ=\sigma$, there holds

$$a(\pi,\sigma^\circ)\leqslant -a(\bar{\pi},\sigma),$$

and this tells us that $a\{\sigma^\circ\to\sigma\}$ is bounded above. Conversely, to show that, if $a\{\sigma^\circ\to\sigma\}$ is bounded above for every σ in Σ, then a has the Clausius property at σ°, we assume that a does *not* have the Clausius property at σ° and show that $a\{\sigma^\circ\to\sigma\}$ is not bounded above for any state σ in Σ. Thus, we suppose that there is a process generator π^* such that

$$\rho_{\pi^*}\sigma^\circ=\sigma^\circ \quad \text{and} \quad a(\pi^*,\sigma^\circ)>0. \tag{2.6}$$

Let σ be in Σ (Figure 30) and choose $\bar{\pi}$ so that

$$\rho_{\bar{\pi}}\sigma^\circ=\sigma. \tag{2.7}$$

Therefore, σ° is in $\mathscr{D}(\bar{\pi})\cap\mathscr{R}(\pi^*)$, σ° is in $\mathscr{D}(\bar{\pi}\pi^*)$, and there holds

$$\left.\begin{aligned} &\rho_{\bar{\pi}\pi^*}\sigma^\circ=\rho_{\bar{\pi}}\rho_{\pi^*}\sigma^\circ=\rho_{\bar{\pi}}\sigma^\circ=\sigma \\ \text{and}\quad & \\ &a(\bar{\pi}\pi^*,\sigma^\circ)=a(\pi^*,\sigma^\circ)+a(\bar{\pi},\sigma^\circ). \end{aligned}\right\} \tag{2.8}$$

Because of $(2.8)_1$, the process generator $\bar{\pi}\pi^*$ also satisfies (2.7), and this permits us to replace $\bar{\pi}$ by $\bar{\pi}\pi^*$ in (2.8):

$$\left.\begin{aligned} &\rho_{(\bar{\pi}\pi^*)\pi^*}\sigma^\circ=\sigma \\ &a((\bar{\pi}\pi^*)\pi^*,\sigma^\circ)=a(\pi^*,\sigma^\circ)+a(\bar{\pi}\pi^*,\sigma^\circ) \\ &\qquad =2a(\pi^*,\sigma^\circ)+a(\bar{\pi},\sigma^\circ). \end{aligned}\right\} \tag{2.9}$$

It should now be clear that we may precede $\bar{\pi}$ by an arbitrary number n of repetitions of π^* and the resulting process generator $\pi^{(n)}$ obeys the relations

$$\left.\begin{aligned} &\rho_{\pi^{(n)}}\sigma^\circ=\sigma \\ &a(\pi^{(n)},\sigma^\circ)=\mathrm{n}\,a(\pi^*,\sigma^\circ)+a(\bar{\pi},\sigma^\circ). \end{aligned}\right\} \tag{2.10}$$

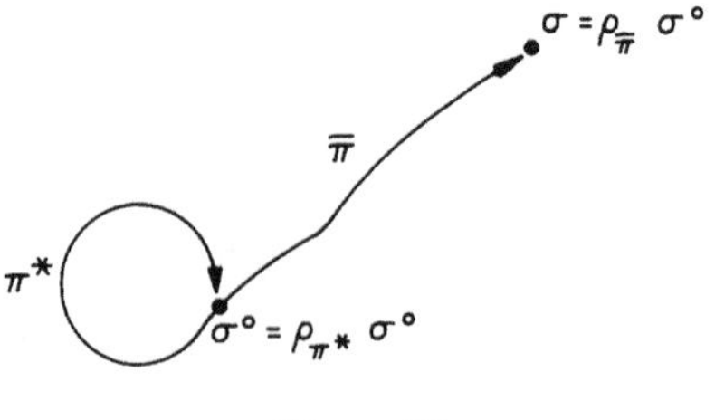

Figure 30.

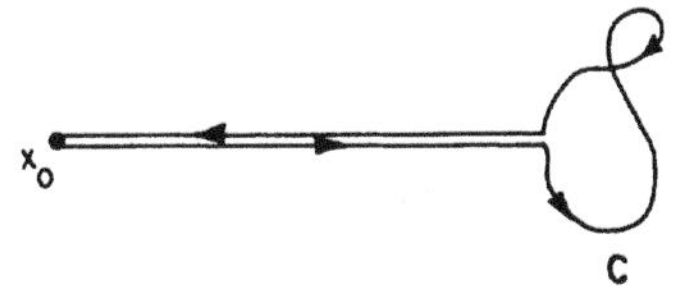

Figure 31.

According to (2.6), $a(\pi^*, \sigma^\circ)$ is positive, and (2.10) and (2.4) then tell us that $a\{\sigma^\circ \to \sigma\}$ is not bounded above. □

The next lemma has a counterpart in the classical theory of line integrals: if the line integral of a vector field vanishes for all closed curves passing through a given point x_0, then the line integral vanishes on all closed curves. Figure 31 indicates why this is so by showing how an arbitrary closed curve C determines a closed curve through x_0 which yields the same value for the line integral.

Lemma 2.2. *a has the Clausius property at a state σ° if and only if a has the Clausius property at every state in Σ.*

PROOF. One implication in the statement of the lemma is clear. To prove its converse, suppose that a has the Clausius property at σ° and let σ be in Σ. If we can show that, for every σ' in Σ, the set

$$a\{\sigma \to \sigma'\} := \{a(\pi, \sigma) \mid \rho_\pi \sigma = \sigma'\}$$

is bounded above, then Lemma 2.1 will imply that a has the Clausius property at σ. According to (S1), Definition 1.1, Chapter II, there is a process generator $\bar{\pi}$ such that $\rho_{\bar{\pi}}\sigma^\circ = \sigma$. For each state σ' in Σ and each process generator π for which $\rho_\pi \sigma = \sigma'$, it follows that $\pi\bar{\pi}$ is defined (Figure 32), $\rho_{\pi\bar{\pi}}\sigma^\circ = \sigma'$, and

$$a(\pi, \sigma) = a(\pi\bar{\pi}, \sigma^\circ) - a(\bar{\pi}, \sigma^\circ). \tag{2.11}$$

Because a has the Clausius property at σ°, $a\{\sigma^\circ \to \sigma'\}$ is bounded above (by

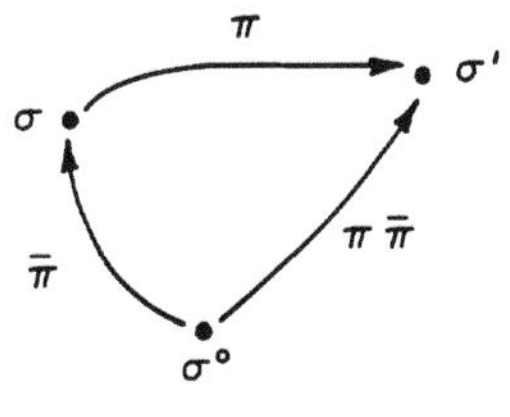

Figure 32.

Lemma 2.1). Therefore, there is a positive number M such that

$$a(\pi\bar{\pi}, \sigma^\circ) < M \tag{2.12}$$

for *all* process generators π as above, i.e., for which $\rho_\pi \sigma = \sigma'$. Relations (2.11) and (2.12) show that

$$a(\pi, \sigma) < M - a(\bar{\pi}, \sigma^\circ)$$

for all such π, so that $a\{\sigma \to \sigma'\}$ is bounded above. □

Lemma 2.3. *If a has the Clausius property at a state σ°, then for every pair of states σ_1, σ_2 in Σ, the set*

$$a\{\sigma_1 \to \sigma_2\} := \{a(\pi, \sigma_1) \mid \rho_\pi \sigma_1 = \sigma_2\} \tag{2.13}$$

is bounded above, non-empty, and, hence, the number

$$s(\sigma_1, \sigma_2) = \sup a\{\sigma_1 \to \sigma_2\} \tag{2.14}$$

is well defined.

PROOF. If a has the Clausius property at σ°, then Lemma 2.2 implies that a has the Clausius property at every state σ_1, and Lemma 2.1 tells us that $a\{\sigma_1 \to \sigma_2\}$ is bounded above for every pair of states σ_1, σ_2 in Σ. By (S1), the set $a(\sigma_1 \to \sigma_2\}$ is non-empty, for there is at least one process generator π with $\rho_\pi \sigma_1 = \sigma_2$ and, hence, $a(\pi, \sigma_1)$ is in the set $a\{\sigma_1 \to \sigma_2\}$. Because $a\{\sigma_1 \to \sigma_2\} \subset \mathbb{R}$ is non-empty and bounded above, the supremum of this set is well defined. □

The final in our series of lemmas gives a result reminiscent of the "triangle inequality" (Figure 33).

Lemma 2.4. *If a has the Clausius property at a state in Σ and if s is defined as in* (2.14), *then for every $\sigma_1, \sigma_2, \sigma_3$ in Σ there holds*

$$s(\sigma_1, \sigma_2) + s(\sigma_2, \sigma_3) \leqslant s(\sigma_1, \sigma_3). \tag{2.15}$$

PROOF. Let $\sigma_1, \sigma_2, \sigma_3$ be in Σ and let ε be a positive number. By the definition of *supremum*, there exist process generators π_{12}^ε and π_{23}^ε (Figure 34) such that

$$\rho_{\pi_{12}^\varepsilon} \sigma_1 = \sigma_2, \qquad \rho_{\pi_{23}^\varepsilon} \sigma_2 = \sigma_3,$$

$$s(\sigma_1, \sigma_2) - \frac{\varepsilon}{2} < a(\pi_{12}^\varepsilon, \sigma_1),$$

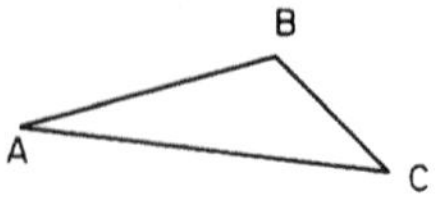

$\overline{AC} \leq \overline{AB} + \overline{BC}$

Figure 33.

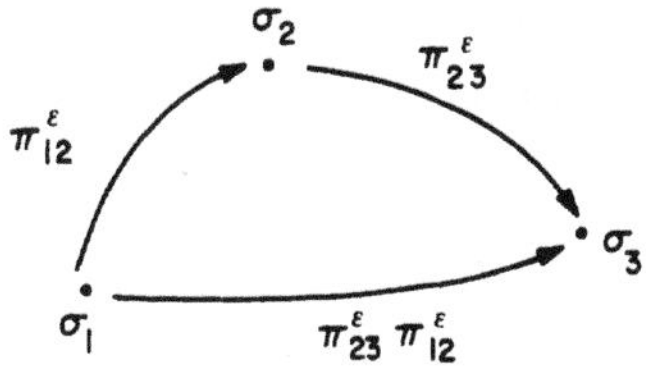

Figure 34.

and

$$s(\sigma_2,\sigma_3)-\frac{\varepsilon}{2}<a(\pi_{23}^{\varepsilon},\sigma_2).$$

These relations and Definition 2.1 of Chapter II yield the inequality

$$s(\sigma_1,\sigma_2)+s(\sigma_2,\sigma_3)-\varepsilon<a(\pi_{12}^{\varepsilon},\sigma_1)+a(\pi_{23}^{\varepsilon},\rho_{\pi_{12}^{\varepsilon}}\sigma_1)$$
$$=a(\pi_{23}^{\varepsilon}\pi_{12}^{\varepsilon},\sigma_1)\leqslant s(\sigma_1,\sigma_3),$$

and the fact that this relation holds for arbitrary positive numbers ε implies (2.15). □

In order to state our main result about actions with the Clausius property, we need an additional concept.

Definition 2.2. A function A: $\Sigma\to\mathbb{R}$ is said to be an *upper potential* for an action a if, for every pair of states σ_1,σ_2 in Σ and every process generator π with $\rho_\pi\sigma_1=\sigma_2$ (Figure 35), there holds

$$a(\pi,\sigma_1)\leqslant A(\sigma_2)-A(\sigma_1);\tag{2.16}$$

in other words, for every process $(\pi,\sigma)\in\Pi\diamondsuit\Sigma$,

$$a(\pi,\sigma)\leqslant A(\rho_\pi\sigma)-A(\sigma).\tag{2.17}$$

Theorem 2.1. *An action a has the Clausius property at a state σ° if and only if a has an upper potential; in fact, the function A°: $\Sigma\to\mathbb{R}$ defined by*

$$A^\circ(\sigma)=s(\sigma^\circ,\sigma)\tag{2.18}$$

is an upper potential for a.

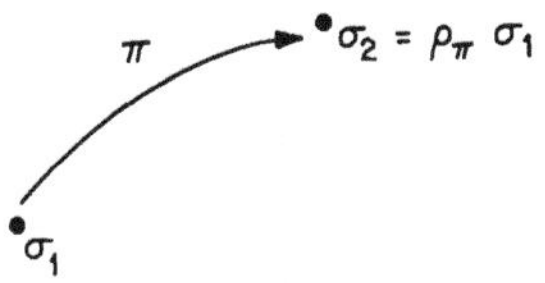

Figure 35.

Proof. Suppose a has the Clausius property at σ°, and note, by Lemma 2.3, that A° is well defined. For each $\sigma_1, \sigma_2 \in \Sigma$ and $\pi \in \Pi$ such that $\rho_\pi\sigma_1 = \sigma_2$, there holds, by (2.14) and (2.15),

$$\begin{aligned} a(\pi,\sigma_1) &\leqslant s(\sigma_1,\sigma_2) \leqslant s(\sigma^\circ,\sigma_2) - s(\sigma^\circ,\sigma_1) \\ &= A^\circ(\sigma_2) - A^\circ(\sigma_1), \end{aligned}$$

which shows that A° is an upper potential for a. Conversely, if a has an upper potential A, then, for each *cycle* (π, σ°), the relation (2.17) yields

$$a(\pi,\sigma^\circ) \leqslant A(\rho_\pi\sigma^\circ) - A(\sigma^\circ) = A(\sigma^\circ) - A(\sigma^\circ) = 0,$$

i.e., a has the Clausius property at σ°. □

This theorem provides a condition which does not mention cycles and which is equivalent to the Clausius property at a state. In fact, the existence of a function A on Σ satisfying (2.17) provides a condition which applies to *every* process of (Σ, Π), not just to cycles.

The main result of this section is an application of Theorem 2.1 to the thermodynamical systems studied in Chapter IV. The hypothesis of this result is precisely that of the Accumulation Theorem (Theorem 6.1 of Chapter IV).

Corollary 2.1. *Let $\mathscr{S}$ be a thermodynamical system, let $\mathscr{G}$ be an ideal gas, and suppose that $\mathscr{S}$ and $\mathscr{G}$ preserve the Second Law with respect to the product operation. It follows that $\mathscr{S}$ obeys the Second Law if and only if there is a real-valued function* S *on the state space Σ of $\mathscr{S}$ which satisfies*

$$\int_0^\infty H_{\mathscr{S}}\big(\pi,\sigma,\varphi_{\mathscr{G}}^{-1}(\theta)\big)\theta^{-2}\,d\theta \leqslant \mathrm{S}(\rho_\pi\sigma) - \mathrm{S}(\sigma) \tag{2.19}$$

for every process (π,σ) of $\mathscr{S}$.

Proof. For each θ in $\mathbb{R}^{++}$, $(\pi,\sigma) \mapsto H_{\mathscr{S}}(\pi,\sigma,\varphi_{\mathscr{G}}^{-1}(\theta))$ is an action for (Σ, Π) (by (TS3) of Definition 4.1, Chapter IV), and it follows that the formula

$$\mathrm{I}_{\mathscr{S}}(\pi,\sigma) := \int_0^\infty H_{\mathscr{S}}\big(\pi,\sigma,\varphi_{\mathscr{G}}^{-1}(\theta)\big)\theta^{-2}\,d\theta \tag{2.20}$$

defines an action $\mathrm{I}_{\mathscr{S}}$ for (Σ, Π). The hypothesis of the corollary and the Accumulation Theorem tell us that the Second Law is equivalent to the Accumulation Inequality, (6.2) of Chapter IV, which by (2.3) is equivalent to the statement that $\mathrm{I}_{\mathscr{S}}$ has the Clausius property at every state of $\mathscr{S}$. The corollary now follows immediately from Theorem 2.1. □

A function S which obeys (2.19), i.e., an upper potential for the action $\mathrm{I}_{\mathscr{S}}$ in (2.20), is called an *entropy function* for $\mathscr{S}$, because of the similarity between (2.19) and the classical "Clausius-Planck inequality":

$$\int \frac{dQ}{T} \leqslant \Delta \mathrm{S}.$$

This relation is written in the traditional notation of classical thermodynamics; in our notation for homogeneous fluid bodies it reads

$$\int_{\Gamma(\pi_t,\sigma)} \frac{\tilde{\lambda}}{\theta} dV + \frac{\kappa}{\theta} d\theta \leqslant \mathrm{S}(\rho_{\pi_t}\sigma) - \mathrm{S}(\sigma),$$

which turns out to be equivalent to (3.28) of Chapter I when it applies to all processes (π_t, σ) of a homogeneous fluid body. Because the symbol S in the Clausius–Planck inequality is called "the entropy," we feel justified in using the term "entropy function" for the upper potential S in (2.19).

In summary, when the accumulation integral $\mathrm{I}_{\mathscr{S}}$ for a thermodynamical system is also an action with the Clausius property, then an upper potential for $\mathrm{I}_{\mathscr{S}}$ is called an entropy function for $\mathscr{S}$.

The remainder of this section is devoted to a detailed study of upper potentials for actions. In this study, we continue to work in the context of systems with perfect accessibility and appeal to the concept of a thermodynamical system only to interpret some of the results obtained. We begin by giving some elementary properties of upper potentials.

(UP1) If A is an upper potential for an action $\mathscr{a}$ and if (π, σ) is a process for which $\mathscr{a}(\pi, \sigma) = 0$, then $A(\rho_\pi \sigma) \geqslant A(\sigma)$.

(UP2) If A is an upper potential for an action $\mathscr{a}$, then so is $A + c$ with c any constant. If A_1 and A_2 are upper potentials for $\mathscr{a}$ and α is in $[0,1]$, then $\alpha A_1 + (1-\alpha) A_2$ is an upper potential for $\mathscr{a}$.

(UP3) If A is an upper potential for an action $\mathscr{a}$, and (π, σ) is a process such that

$$A(\rho_\pi \sigma) - A(\sigma) = \mathscr{a}(\pi, \sigma),$$

then $\mathscr{a}(\pi, \sigma) = s(\sigma, \rho_\pi \sigma)$.

(UP1) is a generalization of a familiar statement from classical thermodynamics: the entropy of an isolated system cannot decrease. For a thermodynamical system $\mathscr{S}$ which obeys (2.19), a corresponding result is the assertion that entropy cannot decrease in a process (π, σ) whose accumulation function $H_{\mathscr{S}}(\pi, \sigma, \cdot)$ is zero. The first part of (UP2) is a result which also is valid for potentials of vector fields and gives a trivial type of non-uniqueness for entropy functions: a constant function plus an entropy function is again an entropy function. The second part of (UP2) tells us that every "convex combination"

$$\alpha A_1 + (1-\alpha) A_2$$

of upper potentials for a given action is again an upper potential, and this suggests that the existence of more than one upper potential implies the existence of infinitely many upper potentials (Figure 36).

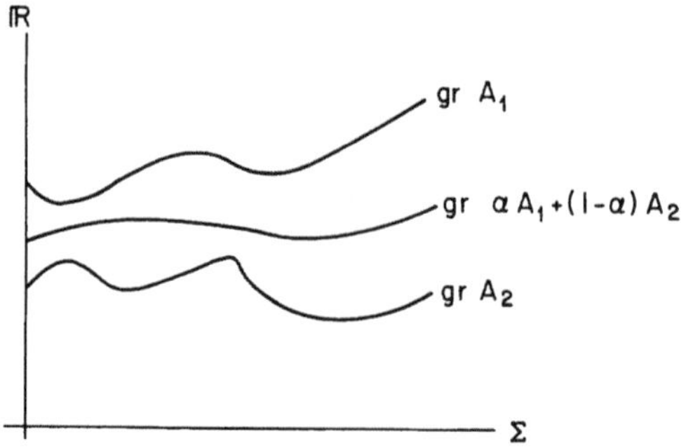

Figure 36.

Of course, (UP2) tells us that the set $\mathscr{U}(a)$ defined by

$$\mathscr{U}(a)=\{A:\Sigma\to\mathbb{R}\,|\,A \text{ is an upper potential for } a\} \tag{2.21}$$

is a convex subset of the set of all functions from Σ into $\mathbb{R}$.

We turn now to a more detailed description of the following subset of $\mathscr{U}(a)$:

$$\mathscr{U}^\circ=\mathscr{U}^\circ(a):=\{A\in\mathscr{U}(a)\,|\,A(\sigma^\circ)=0\} \tag{2.22}$$

with σ° a preassigned state in Σ. By restricting our attention to $\mathscr{U}^\circ$, we remove the possibility that two distinct upper potentials for a can differ by a constant.

Theorem 2.2. *Let a have the Clausius property at σ°, let s be given by (2.14), and suppose that there is a process generator $\bar{\pi}$ for which*

$$\rho_{\bar{\pi}}\sigma^\circ=\sigma^\circ \quad \text{and} \quad a(\bar{\pi},\sigma^\circ)=0. \tag{2.23}$$

It follows that the functions A° and $\hat{A}^\circ$ defined by

$$\left.\begin{aligned} A^\circ(\sigma)&=s(\sigma^\circ,\sigma)\\ \hat{A}^\circ(\sigma)&=-s(\sigma,\sigma^\circ)\end{aligned}\right\} \tag{2.24}$$

are in $\mathscr{U}^\circ$. In fact, A° and $\hat{A}^\circ$ are the smallest and largest elements of $\mathscr{U}^\circ$ in the sense that

$$A^\circ(\sigma)\leqslant A(\sigma)\leqslant\hat{A}^\circ(\sigma) \tag{2.25}$$

for every σ in Σ and A in $\mathscr{U}^\circ$. Therefore, the set $\mathscr{U}^\circ$ has exactly one element if and only if $A^\circ=\hat{A}^\circ$, i.e., a has exactly one upper potential which vanishes at σ° if and only if

$$s(\sigma,\sigma^\circ)=-s(\sigma^\circ,\sigma) \tag{2.26}$$

for every σ in Σ.

Proof. We note first that (2.23) implies that $s(\sigma^\circ,\sigma^\circ)=0$. In fact, there holds

$$0\geqslant s(\sigma^\circ,\sigma^\circ)\geqslant a(\bar{\pi},\sigma^\circ)=0;$$

the first inequality follows from the Clausius property at σ° and the second from the fact that $a(\tilde{\pi}, \sigma^\circ)$ is in the set $a\{\sigma^\circ \to \sigma^\circ\}$. It follows from this and Theorem 2.1 that A° is in $\mathscr{U}^\circ$. To show that $\hat{A}^\circ$ is in $\mathscr{U}^\circ$, we need only show that $\hat{A}^\circ$ is an upper potential for a. Let σ_1, σ_2 be in Σ and $\pi \in \Pi$ be such that $\rho_\pi \sigma_1 = \sigma_2$. By (2.15), we have

$$-s(\sigma_2, \sigma^\circ) - (-s(\sigma_1, \sigma^\circ)) = s(\sigma_1, \sigma^\circ) - s(\sigma_2, \sigma^\circ)$$
$$\geqslant s(\sigma_1, \sigma_2),$$

so that

$$\hat{A}^\circ(\sigma_2) - \hat{A}^\circ(\sigma_1) \geqslant s(\sigma_1, \sigma_2) \geqslant a(\pi, \sigma_1).$$

Therefore, $\hat{A}^\circ$ is an upper potential for a. Now, if σ is in Σ, A is in $\mathscr{U}^\circ$, and $\rho_\pi \sigma^\circ = \sigma$, then

$$A(\sigma) = A(\sigma) - A(\sigma^\circ) \geqslant a(\pi, \sigma^\circ),$$

so that $A(\sigma)$ is an upper bound for $a\{\sigma^\circ \to \sigma\}$. By the definition of *supremum*, we have

$$A(\sigma) \geqslant \sup a\{\sigma^\circ \to \sigma\} = s(\sigma^\circ, \sigma) = A^\circ(\sigma). \tag{2.27}$$

Similarly, if $\rho_{\tilde{\pi}} \sigma = \sigma^\circ$, then

$$-A(\sigma) = A(\sigma^\circ) - A(\sigma) \geqslant a(\tilde{\pi}, \sigma),$$

$-A(\sigma)$ is an upper bound for $a\{\sigma \to \sigma^\circ\}$, and

$$-A(\sigma) \geqslant s(\sigma, \sigma^\circ) = -\hat{A}^\circ(\sigma). \tag{2.28}$$

Relations (2.27) and (2.28) yield (2.25), and this completes the proof of Theorem 2.2. □

Relation (2.26) merits further discussion, because it provides a criterion for determining whether or not a thermodynamical system has exactly one entropy function normalized to vanish at a given state. If we note that the first argument of the function s in (2.14) plays the role of an initial state and the second argument represents a final state, then it is clear that (2.26) is of the form: a quantity associated with an action a changes sign when initial and final states are interchanged. We recall that in classical thermodynamics a condition of "reversibility" is satisfied, in that quantities such as $H(\mathbb{P})$ and $W(\mathbb{P})$ change sign whenever a path $\mathbb{P}$ is replaced by its reversal $\mathbb{P}_r$. This suggests that (2.26) might be valid for special systems whose processes possess reversals on which the action a changes sign. Although this assertion can be justified by precise arguments, it is important to remember that processes of general systems with perfect accessibility are not required to be "reversible." In fact, any reasonable formulation of the concept of "reversibility," if imposed on all processes of a system, would remove from the scope of the modern treatment important examples such as the viscous and the elastic-plastic filaments of Chapters VI and VII. Therefore, (2.26) embodies a condition weaker than "reversibility." The next result makes this condition more precise.

Theorem 2.3. *Let* a *and* σ° *satisfy the hypothesis of Theorem 2.2. There is exactly one upper potential for* a *which vanishes at* σ° *if and only if, for each* $\varepsilon > 0$ *and state* σ, *there are process generators* π_ε *and* $\hat{\pi}_\varepsilon$ *such that*

$$\rho_{\pi_\varepsilon}\sigma^\circ = \sigma, \qquad \rho_{\hat{\pi}_\varepsilon}\sigma = \sigma^\circ, \tag{2.29}$$

and

$$-\varepsilon < a(\pi_\varepsilon, \sigma^\circ) + a(\hat{\pi}_\varepsilon, \sigma) \leqslant 0; \tag{2.30}$$

the upper potential for a *is then given by*

$$A^\circ(\sigma) = s(\sigma^\circ, \sigma) = \lim_{\varepsilon \to 0} a(\pi_\varepsilon, \sigma^\circ). \tag{2.31}$$

PROOF. Suppose first that a has exactly one normalized upper potential, so that (2.26) is valid for every state σ. For each $\varepsilon > 0$ and state σ, the definition (2.14) of s tells us that there are process generators π_ε and $\hat{\pi}_\varepsilon$ such that (2.29) holds and

$$s(\sigma^\circ, \sigma) - \frac{\varepsilon}{2} < a(\pi_\varepsilon, \sigma^\circ) \leqslant s(\sigma^\circ, \sigma), \tag{2.32}$$

$$s(\sigma, \sigma^\circ) - \frac{\varepsilon}{2} < a(\hat{\pi}_\varepsilon, \sigma) \leqslant s(\sigma, \sigma^\circ). \tag{2.33}$$

It is clear that (2.26), (2.32), and (2.33) imply (2.30). Moreover, (2.31) follows immediately from (2.32) if we let ε tend to zero in that relation. Conversely, if for each ε and σ there are process generators satisfying (2.29) and (2.30), then (2.14) and these relations yield

$$-s(\sigma, \sigma^\circ) - \varepsilon \leqslant -a(\hat{\pi}_\varepsilon, \sigma) - \varepsilon < a(\pi_\varepsilon, \sigma^\circ) \leqslant s(\sigma^\circ, \sigma); \tag{2.34}$$

because $\varepsilon > 0$ is arbitrary, (2.34) implies

$$-s(\sigma, \sigma^\circ) \leqslant s(\sigma^\circ, \sigma). \tag{2.35}$$

Of course, the relation $s(\sigma^\circ, \sigma^\circ) = 0$ (which simply amounts to the fact that $A^\circ(\sigma^\circ)$ is zero) and the inequality (2.15) (with $\sigma_1 = \sigma_3 = \sigma^\circ$ and $\sigma_2 = \sigma$) imply that

$$s(\sigma, \sigma^\circ) + s(\sigma^\circ, \sigma) \leqslant 0. \tag{2.36}$$

Relations (2.35) and (2.36) yield (2.26) forthwith. □

We can write (2.30) in the following form:

$$a(\hat{\pi}_\varepsilon, \sigma) = -a(\pi_\varepsilon, \sigma^\circ) + o_\varepsilon(1), \tag{2.37}$$

where $o_\varepsilon(1)$ tends to zero as ε tends to zero. Relations (2.37) and (2.29) are equivalent to (2.26), and they permit us to interpret (2.26) in the following manner: *for every* $\varepsilon > 0$ *and state* σ *there are processes* $(\pi_\varepsilon, \sigma^\circ)$ *and* $(\hat{\pi}_\varepsilon, \sigma)$

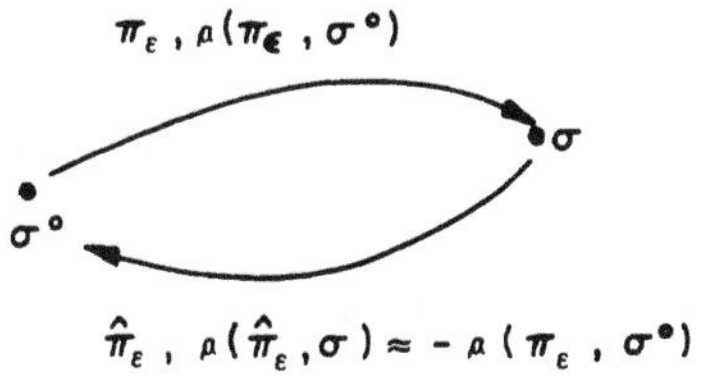

Figure 37.

which take the system from σ° to σ and then return it to σ° and which, to within terms of order ε, remove as much of the quantity a in returning the system to σ° as was added in bringing it to σ. This suggests that we interpret (2.26) as a *condition of (approximate) restorability from σ°* (Figure 37). It differs from "reversibility" in that there need be no special relation between $(\pi_\varepsilon, \sigma^\circ)$ and $(\hat{\pi}_\varepsilon, \sigma)$ in terms of some notion of "reversal" of processes. However, when (Σ, Π) represents a homogeneous fluid body and a is the action

$$a(\pi_t, \sigma) = \int_{\Gamma(\pi_t, \sigma)} \frac{\tilde{\lambda}}{\theta} dV + \frac{\kappa}{\theta} d\theta,$$

then (2.37) *is* valid with $(\hat{\pi}_\varepsilon, \sigma)$ corresponding to the reversal of $(\pi_\varepsilon, \sigma^\circ)$ and with $o_\varepsilon(1) = 0$, and we obtain the result already contained in Chapter I: a homogeneous fluid body which obeys the laws of thermodynamics has exactly one entropy function normalized to vanish at a preassigned state.

It is useful to apply Theorem 2.3 in the case where (Σ, Π) is associated with a thermodynamical system $\mathscr{S}$ and a is the accumulation integral $I_{\mathscr{S}}$ of (2.20).

Corollary 2.2. *Let a thermodynamical system $\mathscr{S}$ satisfy the Accumulation Inequality*

$$(\pi, \sigma) \in (\Pi_{\mathscr{S}} \diamond \Sigma_{\mathscr{S}})_{\text{cyc}} \Rightarrow I_{\mathscr{S}}(\pi, \sigma) \leqslant 0, \tag{2.38}$$

where $I_{\mathscr{S}}$ is the accumulation integral

$$I_{\mathscr{S}}(\pi, \sigma) = \int_0^\infty H_{\mathscr{S}}(\pi, \sigma, \varphi_{\mathscr{G}}^{-1}(\theta))\theta^{-2} d\theta$$

and $\mathscr{G}$ is a given ideal gas, and let σ° be a state with the following properties:

(1) *there is a process generator $\bar{\pi}$ for which $(\bar{\pi}, \sigma^\circ)$ is a cycle and $I_{\mathscr{S}}(\bar{\pi}, \sigma^\circ) = 0$;*
(2) *for each $\varepsilon > 0$ and $\sigma \in \Sigma_{\mathscr{S}}$ there are process generators π_ε and $\hat{\pi}_\varepsilon$ such that $\rho_{\pi_\varepsilon}\sigma^\circ = \sigma$, $\rho_{\hat{\pi}_\varepsilon}\sigma = \sigma^\circ$, and*

$$-\varepsilon < I_{\mathscr{S}}(\pi_\varepsilon, \sigma^\circ) + I_{\mathscr{S}}(\hat{\pi}_\varepsilon, \sigma) \leqslant 0. \tag{2.39}$$

The system $\mathscr{S}$ then has exactly one entropy function S° which vanishes at σ°;

this function is given by the formulae

$$S^\circ(\sigma) = \lim_{\varepsilon \to 0} \int_0^\infty H_{\mathscr{S}}\big(\pi_\varepsilon, \sigma^\circ, \varphi_{\mathscr{G}}^{-1}(\theta)\big)\theta^{-2}\, d\theta$$

$$= -\lim_{\varepsilon \to 0} \int_0^\infty H_{\mathscr{S}}\big(\hat{\pi}_\varepsilon, \sigma, \varphi_{\mathscr{G}}^{-1}(\theta)\big)\theta^{-2}\, d\theta. \tag{2.40}$$

3. Actions with the Conservation Property

In our introduction to the concept of an action with the Clausius property, we observed that the Second Law for homogeneous fluid bodies and for thermodynamical systems both are equivalent to the assertion that a function on cycles have non-positive values. In the first case the function represents the line integral of heat gained divided by temperature and, in the second, the accumulation integral. A passing glance at our study of both classical and modern versions of the First Law reveals a similar situation, except that the corresponding functions *vanish* on cycles. In fact, Theorem 3.1 of Chapter 3 tells us that the First Law for thermodynamical systems is equivalent to Joule's relation, which may be written in the form

$$(\pi, \sigma) \in (\Pi_{\mathscr{S}} \diamondsuit \Sigma_{\mathscr{S}})_{\text{cyc}} \Rightarrow \frac{R}{\lambda} H_{\mathscr{S}}(\pi, \sigma) - W_{\mathscr{S}}(\pi, \sigma) = 0.$$

Because $W_{\mathscr{S}}$ and $H_{\mathscr{S}}$ are actions for $(\Sigma_{\mathscr{S}}, \Pi_{\mathscr{S}})$, it follows that so is $(R/\lambda)H_{\mathscr{S}} - W_{\mathscr{S}}$, so that Joule's relation is an assertion that a given action for a system vanish on every cycle. This suggests the following definition.

Definition 3.1. An action a for a system with perfect accessibility (Σ, Π) has the *conservation property at a state* σ° if there holds

$$(\pi, \sigma^\circ) \in (\Pi \diamondsuit \Sigma)_{\text{cyc}} \Rightarrow a(\pi, \sigma^\circ) = 0. \tag{3.1}$$

This terminology comes from the classical theory of line integrals of vector fields, in which a vector field is said to be conservative if its line integral vanishes on closed curves.

An elementary fact about actions with the conservation property turns out to be useful in our study of these actions: *a has the conservation property at σ° if and only if both a and $-a$ have the Clausius property at σ°.* This follows from the fact that $a(\pi, \sigma^\circ) = 0$ holds if and only if both $a(\pi, \sigma^\circ) \leqslant 0$ and $-a(\pi, \sigma^\circ) \leqslant 0$ hold. From this result and Lemma 2.2, we obtain the following result.

Lemma 3.1. *a has the conservation property at σ° if and only if a has the conservation property at every state in Σ.*

The sets $a\{\sigma_1 \to \sigma_2\}$ defined in (2.13) are easily described when a has the conservation property at a state.

Lemma 3.2. *If a has the conservation property at a state σ°, then for every σ_1, σ_2 in Σ the set $a\{\sigma_1 \to \sigma_2\}$ is a singleton. Hence, for every $\pi \in \Pi$ such that $\rho_\pi \sigma_1 = \sigma_2$, there holds*

$$a\{\sigma_1 \to \sigma_2\} = \{a(\pi, \sigma_1)\}$$

$$s(\sigma_1, \sigma_2) := \sup a\{\sigma_1 \to \sigma_2\} = a(\pi, \sigma_1). \tag{3.2}$$

Moreover, for every σ_1, σ_2 in Σ,

$$s(\sigma_2, \sigma_1) = -s(\sigma_1, \sigma_2). \tag{3.3}$$

Proof. By Lemma 3.1, if a has the conservation property at σ°, then it has the conservation property at every state in Σ. For each $\sigma_1, \sigma_2 \in \Sigma$, choose $\tilde{\pi}$ in Π so that $\rho_{\tilde{\pi}} \sigma_2 = \sigma_1$. For every π such that $\rho_\pi \sigma_1 = \sigma_2$, the pair $(\tilde{\pi}\pi, \sigma_1)$ is a cycle (Figure 38), and the fact that a has the conservation property at σ_1 implies that

$$0 = a(\tilde{\pi}\pi, \sigma_1) = a(\pi, \sigma_1) + a(\tilde{\pi}, \sigma_2), \tag{3.4}$$

i.e., $a(\pi, \sigma_1) = -a(\tilde{\pi}, \sigma_2)$ for *every* π as above. This yields both assertions in (3.2), and (3.3) then follows immediately from (3.2) and (3.4). □

The next result gives a strengthened version of the "reverse triangle inequality" (2.15).

Lemma 3.3. *If a has the conservation property at a state, then there holds*

$$s(\sigma_1, \sigma_2) + s(\sigma_2, \sigma_3) = s(\sigma_1, \sigma_3) \tag{3.5}$$

for every $\sigma_1, \sigma_2, \sigma_3$ in Σ.

Proof. If π_{12} and π_{23} are such that $\rho_{\pi_{12}} \sigma_1 = \sigma_2$ and $\rho_{\pi_{23}} \sigma_2 = \sigma_3$, then $\rho_{\pi_{23}\pi_{12}} \sigma_1 = \sigma_3$ and there holds, by (3.2),

$$\begin{aligned} s(\sigma_1, \sigma_2) + s(\sigma_2, \sigma_3) &= a(\pi_{12}, \sigma_1) + a(\pi_{23}, \sigma_2) \\ &= a(\pi_{23}\pi_{12}, \sigma_1) \\ &= s(\sigma_1, \sigma_3). \end{aligned}$$

□

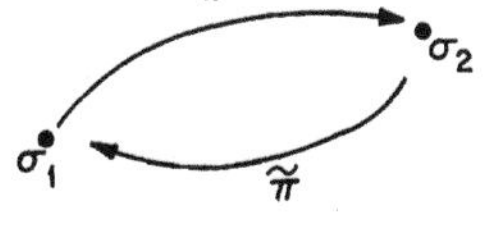

Figure 38.

Our main results on actions with the conservation property are best stated in terms of the notion of a "potential" for an action.

Definition 3.2. A *potential* A for an action a is a function from Σ into $\mathbb{R}$ such that

$$\rho_\pi \sigma_1 = \sigma_2 \Rightarrow A(\sigma_2) - A(\sigma_1) = a(\pi, \sigma_1). \tag{3.6}$$

This concept is an obvious analogue of the notion of potential for a vector field, but it does not require the detailed properties of vector fields and line integrals to be meaningful.

Theorem 3.1. *An action a has the conservation property at a state σ° if and only if a has a potential. In this case, there is exactly one potential for a which vanishes at σ°.*

PROOF. By Lemmas 3.2 and 3.3, for each $\sigma_1, \sigma_2 \in \Sigma$ and process generator π for which $\rho_\pi \sigma_1 = \sigma_2$, there holds

$$s(\sigma^\circ, \sigma_2) - s(\sigma^\circ, \sigma_1) = s(\sigma_1, \sigma_2) = a(\pi, \sigma_1),$$

i.e., $\sigma \mapsto s(\sigma^\circ, \sigma)$ is a potential for a. Conversely, if a has a potential A, then for each cycle (π, σ)

$$a(\pi, \sigma) = A(\rho_\pi \sigma) - A(\sigma) = A(\sigma) - A(\sigma) = 0,$$

i.e., a has the conservation property at every state. If A is a potential which vanishes at σ° and if $\pi \in \Pi$ and $\sigma \in \Sigma$ are such that $\rho_\pi \sigma^\circ = \sigma$, then

$$A(\sigma) = A(\sigma) - A(\sigma^\circ) = a(\pi, \sigma^\circ).$$

In other words, the value of every such potential at σ is the same number $a(\pi, \sigma^\circ)$. Therefore, every pair of potentials which vanish at σ° must agree everywhere on Σ. □

When the First Law is equivalent to a statement that a given action have the conservation property at a state, then it is natural to call the resulting potentials *energy functions*. According to the above theorem, any two energy functions can differ by at most a constant, and there is exactly one energy function which vanishes at a preassigned state.

Theorem 3.1 of Chapter III and Theorem 3.1 here together have the following consequence.

Corollary 3.1. *Let $\mathcal{S}$ be a thermodynamical system, let $\mathcal{G}$ be an ideal gas, and suppose that $\mathcal{S}$ and $\mathcal{G}$ preserve the First Law with respect to the product operation. It follows that $\mathcal{S}$ obeys the First Law if and only if, for each state σ° of $\mathcal{S}$, there is exactly one energy function* E° *for $\mathcal{S}$ which vanishes at σ°, i.e.,* *there is exactly one function* $E^\circ\colon \Sigma \to \mathbb{R}$ *such that* $E^\circ(\sigma^\circ) = 0$ *and*

$$\rho_\pi \sigma_1 = \sigma_2 \Rightarrow E^\circ(\sigma_2) - E^\circ(\sigma_1) = \left(\frac{R}{\lambda}\right) H_{\mathscr{S}}(\pi, \sigma_1) - W_{\mathscr{S}}(\pi, \sigma_1). \quad (3.7)$$

The function E° *is given by the formula*

$$E^\circ(\sigma) = \left(\frac{R}{\lambda}\right) H_{\mathscr{S}}(\pi, \sigma^\circ) - W_{\mathscr{S}}(\pi, \sigma^\circ) \quad (3.8)$$

for any process generator π *such that* $\rho_\pi \sigma^\circ = \sigma$.

4. Actions with the Dissipation Property

Suppose that a thermodynamical system $\mathscr{S}$ satisfies both Joule's relation *and* the Accumulation Inequality, and let (π, σ) be a cycle for $\mathscr{S}$ whose accumulation function $H_{\mathscr{S}}$ has the form

$$H_{\mathscr{S}}\left(\pi, \sigma, \varphi_{\mathscr{G}}^{-1}(\theta)\right) = \begin{cases} 0, & 0 \leqslant \theta < \theta_0 \\ H_{\mathscr{S}}(\pi, \sigma), & \theta_0 \leqslant \theta. \end{cases} \quad (4.1)$$

The Accumulation Inequality then tells us that

$$0 \geqslant \int_0^\infty H_{\mathscr{S}}\left(\pi, \sigma, \varphi_{\mathscr{G}}^{-1}(\theta)\right) \theta^{-2} \, d\theta = \int_{\theta_0}^\infty H_{\mathscr{S}}(\pi, \sigma) \theta^{-2} \, d\theta = H_{\mathscr{S}}(\pi, \sigma) \theta_0^{-1},$$

so that $H_{\mathscr{S}}(\pi, \sigma) \leqslant 0$. If we let $\check{W}_{\mathscr{S}} = -W_{\mathscr{S}}$, i.e., $\check{W}_{\mathscr{S}}$ represents the work done *on* $\mathscr{S}$ by its environment, then Joule's relation yields

$$\check{W}_{\mathscr{S}}(\pi, \sigma) = -W_{\mathscr{S}}(\pi, \sigma) = -\left(\frac{R}{\lambda}\right) H_{\mathscr{S}}(\pi, \sigma) \geqslant 0, \quad (4.2)$$

i.e., *for any cycle* (π, σ) *whose accumulation function has the form* (4.1), *the work done on* $\mathscr{S}$ *is non-negative.* Such cycles may be called "isothermal," although the use of this term differs from its use in classical thermodynamics, and the inequality (4.2) amounts to the assertion that *the work done on* $\mathscr{S}$ *is non-negative in isothermal cycles.* In fact, for systems which obey the First Law and undergo *only* isothermal processes (including all of the examples in Chapter VI), the Accumulation Inequality is equivalent to (4.2), and we take the Second Law for such systems to be the italicized assertion in the previous sentence.

It is natural to rephrase this condition in the language of systems with perfect accessibility. (In doing so, we interpret the cycles of such a system as all being isothermal, even though the use of this term requires that a system with perfect accessibility also be a thermodynamical system.) In view of our discussion of actions with the Clausius property, it is also natural to formulate a concept analogous to that of an upper potential.

Definition 4.1. An action a for a system with perfect accessibility is said to have the *dissipation property* at σ° if

$$(\pi, \sigma^\circ) \in (\Pi \diamondsuit \Sigma)_{\text{cyc}} \Rightarrow a(\pi, \sigma^\circ) \geqslant 0. \quad (4.3)$$

A *lower potential* for an action a is a function A from Σ into $\mathbb{R}$ such that

$$\rho_\pi \sigma_1 = \sigma_2 \Rightarrow a(\pi, \sigma_1) \geqslant A(\sigma_2) - A(\sigma_1). \tag{4.4}$$

Actions with the dissipation property and lower potentials can be studied in a manner analogous to that used to study actions with the Clausius property and upper potentials. The principal results in that study, Theorems 2.1 and 2.2, have counterparts for actions with the dissipation property which we summarize in one theorem.

Theorem 4.1. *An action a has the dissipation property at a state σ° if and only if a has a lower potential. The set $\mathscr{L}^\circ$ of lower potentials for a which vanish at σ° is then a non-empty, convex set. If there exists $\bar{\pi}$ in Π such that $\rho_{\bar{\pi}}\sigma^\circ = \sigma^\circ$ and $a(\bar{\pi}, \sigma^\circ) = 0$, then $\mathscr{L}^\circ$ has a smallest and largest element given by*

$$\sigma \mapsto -\mathrm{n}(\sigma, \sigma^\circ) := -\inf\{a(\pi, \sigma) | \rho_\pi \sigma = \sigma^\circ\},$$
$$\sigma \mapsto \mathrm{n}(\sigma^\circ, \sigma) := \inf\{a(\pi, \sigma^\circ) | \rho_\pi \sigma^\circ = \sigma\}. \tag{4.5}$$

A lower potential for an action with the dissipation property corresponds to a *Helmholtz free energy function* Ψ in classical thermodynamics:

$$\Psi(V, \theta) = \mathrm{E}(V, \theta) - J\theta \mathrm{S}(V, \theta). \tag{4.6}$$

In fact, for an isothermal path $\mathbb{P}$ of a homogeneous fluid body there holds

$$\begin{aligned}\Psi^\cdot = \mathrm{E}^\cdot - J\theta \mathrm{S}^\cdot &= (J\tilde{\lambda} - p)V^\cdot + J\sigma\theta^\cdot \\ &\quad - J\theta\frac{\tilde{\lambda}}{\theta}V^\cdot - J\theta\frac{\sigma}{\theta}\theta^\cdot \\ &= -pV^\cdot,\end{aligned}$$

so that the change in Ψ on an isothermal path $\mathbb{P}$ is the work done *on* the homogeneous fluid body:

$$\check{W}(\mathbb{P}) = \Psi(V_2, \theta_1) - \Psi(V_1, \theta_1) = \Delta\Psi. \tag{4.7}$$

For a thermodynamical system in which isothermal processes correspond to elements of $\Pi \diamondsuit \Sigma$, relation (4.4) would read

$$(\pi, \sigma) \in (\Pi_{\mathscr{S}} \diamondsuit \Pi_{\mathscr{S}})_{\mathrm{isot}} \Rightarrow \check{W}_{\mathscr{S}}(\pi, \sigma) \geqslant \Delta\Psi, \tag{4.8}$$

where $(\Pi_{\mathscr{S}} \diamondsuit \Sigma_{\mathscr{S}})_{\mathrm{isot}}$ denotes the set of all isothermal processes of $\mathscr{S}$. Thus, the work done on $\mathscr{S}$ in isothermal processes is bounded below by the change in the lower potential Ψ. Because of the similarity between the relations (4.7) and (4.8), we call each lower potential for $\check{W}_{\mathscr{S}}$ a *Helmholtz free energy function*. Thus, a Helmholtz free energy function permits one to obtain a lower bound for the amount of work which can be done on a system in isothermal processes which connect a pair of states.

CHAPTER VI

Isothermal Processes of Homogeneous Filaments

1. Introduction

The approach to the First and Second Laws given in Chapters II through V is an abstract one, axiomatic in nature, and permits us to concentrate on those features of physical systems which are central to the formulation and analysis of these laws. The purpose of this chapter is to introduce elementary examples of systems whose state spaces are not of the type prescribed for homogeneous fluid bodies and yet whose structures are sufficiently simple to yield explicit results without undue effort. To simplify matters as much as possible, we restrict our attention to isothermal processes of one-dimensional, continuous bodies, here called *filaments*. (This terminology is convenient but somewhat misleading, because one generally regards a filament as being incapable of supporting compressive forces, whereas we do not wish to make such a restrictive assumption in the present treatment.) Each of the filaments to be studied can be imagined as a thin string or rod with fixed direction, but whose *length* ℓ may vary as an *external force* f parallel to the filament is varied. The length ℓ is required to be a positive number (Figure 39) and enters into the description of the states available to the filament. The force f represents the force which the environment exerts *on* the filament; f is a real number which, if positive, is called a *tensile force* and, if negative, is called a *compressive force*. It is convenient to give a reference length ℓ_0 which is compatible with absence of an external force, i.e., with $f = 0$. The term "homogeneous" will be used for a filament in a sense similar to that employed in our discussion of homogeneous fluid bodies, namely, every segment *AB* of a filament is subjected to the same force as is the entire filament, although this force may change with time for the filament as a whole (Figure 40).

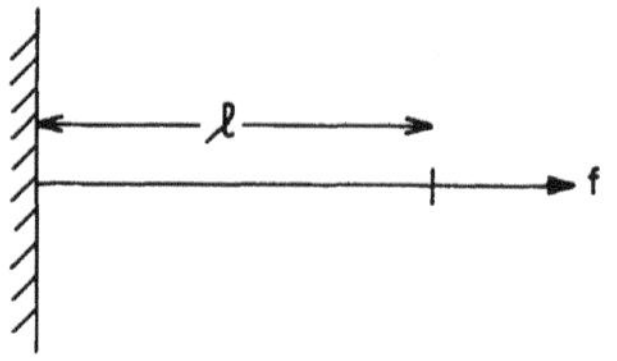

Figure 39.

Figure 40.

We shall discuss several examples of filaments in this chapter. Each will be described as a system with perfect accessibility whose states are lists of the form $(\ell, \ldots)$, so that knowledge of the state determines, among other things, knowledge of the length of the filament. The process generators are functions whose values represent the rate of change of the length of the filament, and the assignments of domains and transformation functions to process generators follow those described in Chapter II, Section 1, for homogeneous fluid bodies. If we assume that Joule's relation holds and that each process of a filament is isothermal, then it is natural, in view of our discussion in Section 4 of Chapter V, to take the Second Law to be the statement that the work done on the filament be non-negative in cycles of the system, i.e., *the work done on the filament is an action with the dissipation property*. In each example, conditions equivalent to the Second Law will be obtained, and the collection of Helmholtz free energy functions introduced for general systems with isothermal processes in Section 4 of Chapter V will be described in detail.

2. Elastic Filaments

The state space Σ is an open interval in $\mathbb{R}^{++}$,

$$\Sigma := (\ell_a, \ell_b) \quad (\text{with } 0 \leqslant \ell_a < \ell_b \leqslant \infty) \tag{2.1}$$

and the set Π of process generators is the set of all piecewise continuous functions of the form π_t: $[0, t) \to \mathbb{R}$ such that, for at least one ℓ in Σ, the oriented curve $\Gamma(\pi_t, \ell)$, parameterized by $\tau \mapsto \ell + \int_0^\tau \pi_t(\xi)\, d\xi$ on $[0, t]$, is contained in Σ. The domain $\mathscr{D}(\pi_t)$ is the set of states ℓ with this property, and the state $\rho_{\pi_t}\ell$ is the final point of $\Gamma(\pi_t, \ell)$. The successive application $\pi_{t''}\pi_{t'}$ of two process generators is the function from $[0, t''+t')$ into $\mathbb{R}$ defined in (1.11), Chapter II; $\pi_{t''}\pi_{t'}$ is in Π whenever there is a state ℓ such that $\Gamma(\pi_{t''}, \rho_{\pi_{t'}}\ell) * \Gamma(\pi_{t'}, \ell)$ is contained in Σ. Each process of (Σ, Π) is a pair (π_t, ℓ) for which $\Gamma(\pi_t, \ell)$ lies in Σ. In fact, we may regard (π_t, ℓ) as a piecewise continuously differentiable parameterization of $\Gamma(\pi_t, \ell)$.

In addition to giving Σ and Π, we give also a continuous function ϕ_e from Σ into $\mathbb{R}$ whose values $\phi_e(\ell)$ represent the external force f on the filament required to maintain it in the state ℓ, i.e., to maintain its length at the value ℓ. The relation

$$f = \phi_e(\ell) \tag{2.2}$$

is called the *equation of state* of the elastic filament (Σ, Π, ϕ_e). The *work done on the filament* in a process (π_t, ℓ) is the number

$$w_e(\pi_t, \ell) := \int_0^t \phi_e(\rho_{\pi_\tau}\ell)\pi_t(\tau)\, d\tau, \tag{2.3}$$

where, for each τ in $(0, t]$, π_τ is the restriction of π_t to $[0, \tau)$ and

$$\rho_{\pi_\tau}\ell = \ell + \int_0^\tau \pi_t(\xi)\, d\xi. \tag{2.4}$$

Because $\rho_{\pi_\tau}\ell$ is the length of the filament at time τ in the process (π_t, ℓ) and

$$\frac{d}{d\tau}(\rho_{\pi_\tau}\ell) = \pi_t(\tau) \tag{2.5}$$

is the rate of change of length at time τ, the integrand appearing in (2.3) is the rate at which the external force does work on the filament, and this justifies our terminology for $w_e(\pi_t, \ell)$. Moreover, if Ψ is an antiderivative of ϕ_e, i.e., $\phi_e(\ell) = \Psi'(\ell)$ for all ℓ in Σ, then (2.3) and (2.5) imply

$$w_e(\pi_t, \ell) = \int_0^t \frac{d}{d\tau}\Psi(\rho_{\pi_\tau}\ell)\, d\tau = \Psi(\rho_{\pi_t}\ell) - \Psi(\ell), \tag{2.6}$$

which yields the following result: *the function w_e is an action with the conservation property at every state, and the Second Law is satisfied for every choice of the continuous function ϕ_e. Moreover, there is exactly one Helmholtz free energy function which vanishes at the state ℓ_0; this function is the antiderivative Ψ° of ϕ_e which vanishes at ℓ_0.*

The simplest example of an elastic filament is that of a "(linear) spring," one for which ϕ_e is given by

$$\phi_e(\ell) = k(\ell - \ell_0), \qquad \ell_a < \ell < \ell_b, \tag{2.7}$$

with k a positive constant. In this case the function Ψ° is given by

$$\Psi^\circ(\ell) = \frac{k}{2}(\ell - \ell_0)^2, \tag{2.8}$$

and $\Psi^\circ(\ell)$ is sometimes called the "stored energy" or "potential energy" of the spring in the state ℓ.

3. Viscous Filaments

Here we take the state space Σ to be the set

$$\begin{aligned} \Sigma := \{(\ell, r) | \ell_a < \ell < \ell_b \quad \text{and} \quad -\infty < r < \infty\} \\ = (\ell_a, \ell_b) \times \mathbb{R}. \end{aligned} \tag{3.1}$$

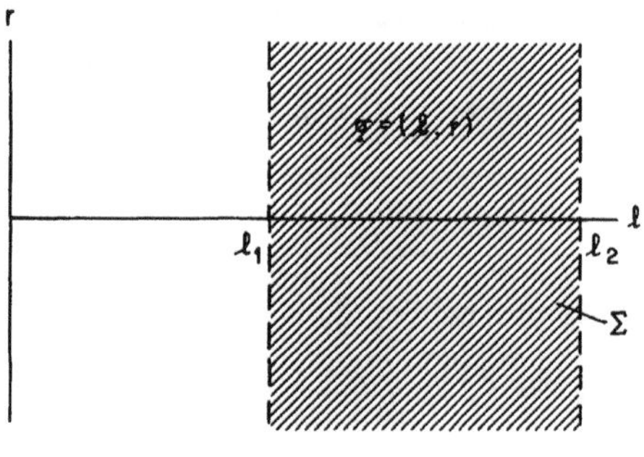

Figure 41.

As we shall see below, the second entry r of a state $\sigma = (\ell, r)$ (Figure 41) represents the rate of change of length of the filament. The set Π is defined exactly as for elastic filaments, as are the sets $\mathscr{D}(\pi_t)$ for π_t in Π. However, the state transformation function ρ_{π_t} in the present case is defined by

$$\rho_{\pi_t}(\ell, r) = (\ell(t), \dot{\ell}(t-)), \tag{3.2}$$

where, for each $\tau \in [0, t]$,

$$\ell(\tau) = \ell + \int_0^\tau \pi_t(\xi)\, d\xi. \tag{3.3}$$

At points of continuity of π_t, we have

$$\dot{\ell}(\tau) = \pi_t(\tau). \tag{3.4}$$

Moreover, by definition,

$$\dot{\ell}(t-) = \lim_{\tau \uparrow t} \dot{\ell}(\tau) = \lim_{\tau \uparrow t} \pi_t(\tau) = \pi_t(t-). \tag{3.5}$$

Thus, the final state $\rho_{\pi_t}(\ell, r)$ for a process $(\pi_t, (\ell, r))$ is the pair $(\ell(t), \dot{\ell}(t-))$ consisting of the final length and final rate of change of length in the process (Figure 42).

We note that condition (S1) in the definition of a system with perfect accessibility in Chapter II is satisfied here, because not only can any two

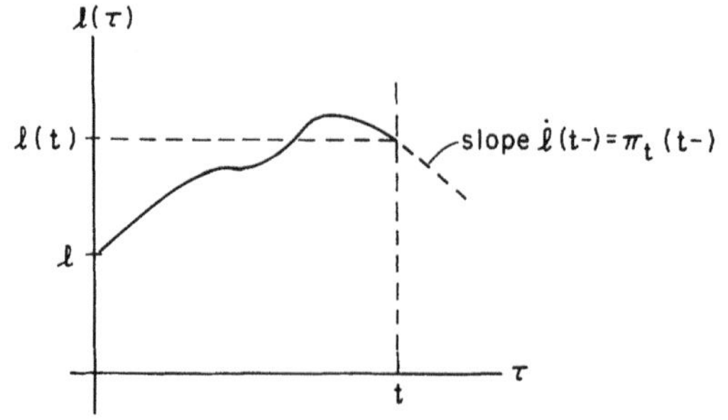

Figure 42.

points in (ℓ_a, ℓ_b) be joined by a curve lying in this interval, but the velocity at the end of the curve can be chosen arbitrarily. Therefore, if one wishes to reach a state (ℓ_2, r_2) starting from (ℓ_1, r_1), then one must traverse a curve $\Gamma(\pi_t, (\ell_1, r_1))$ parameterized by a function $\tau \mapsto \ell(\tau)$ of the form shown in Figure 43. If successive applications of process generators are defined as for elastic filaments, then (Σ, Π) becomes a system with perfect accessibility.

To complete the description of a viscous filament, we must give a function ϕ_v from Σ into $\mathbb{R}$ whose values $\phi_v(\ell, r)$ represent the force on the filament required to maintain it in the state (ℓ, r). In this case, the equation of state takes the form

$$f = \phi_v(\ell, r), \tag{3.6}$$

and we require that the function ϕ_v be continuous on Σ. Relation (3.6) tells us that the force required to maintain the filament in a given state depends upon the length and the most recent rate of change of length of the filament. This is one of the simplest examples of a system whose states depend not only upon present but also past values of various quantities (in our particular case, just one quantity—the length of the filament).

If we write π_τ for the restriction of a process generator π_t to $[0, \tau)$ and $\rho_{\pi_\tau}(\ell, r)$ for the pair $(\ell(\tau), \dot{\ell}(\tau-))$, as in (3.2), then the work done on the filament in $(\pi_t, (\ell, r))$ is defined by

$$\begin{aligned} w_v(\pi_t, (\ell, r)) &= \int_0^t \phi_v\big(\rho_{\pi_\tau}(\ell, r)\big)\pi_t(\tau)\, d\tau \\ &= \int_0^t \phi_v(\ell(\tau), \dot{\ell}(\tau-))\dot{\ell}(\tau-)\, d\tau. \end{aligned} \tag{3.7}$$

It is a routine matter to verify that w_v is an action for (Σ, Π), and *the Second Law here becomes the assertion that w_v has the dissipation property at one and, hence, at every state (ℓ, r)*:

$$\left.\begin{aligned} &\int_0^t \phi_v(\ell(\tau), \dot{\ell}(\tau))\dot{\ell}(\tau)\, d\tau \geqslant 0 \\ &\text{if } \rho_{\pi_t}(\ell, r) = (\ell, r). \end{aligned}\right\} \tag{3.8}$$

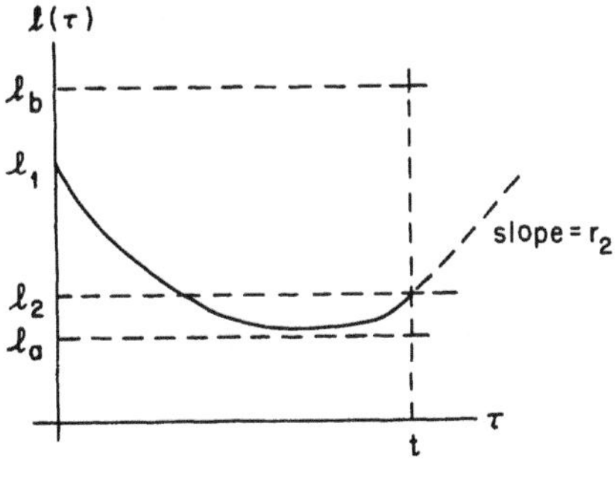

Figure 43.

(We have replaced $\dot{\ell}(\tau -)$ by $\dot{\ell}(\tau)$ in the second argument of ϕ_v in passing from (3.7) to (3.8). This is justified, because $\dot{\ell}$ is defined and continuous at all but a finite number of points of $[0, t]$.)

The relation (3.8) can be rewritten further by observing that

$$\int_0^t \phi_v(\ell(\tau),0)\dot{\ell}(\tau)\,d\tau = 0 \tag{3.9}$$

whenever $\rho_{\pi_t}(\ell, r) = (\ell, r)$. In fact, if we formally replace $\phi_v(\ell(\tau),0)$ by $\phi_e(\ell(\tau))$, then (3.9) follows from our results on elastic filaments. Therefore, (3.8) and (3.9) yield:

$$\left.\begin{aligned}\int_0^t \big[\phi_v(\ell(\tau),\dot{\ell}(\tau)) - \phi_v(\ell(\tau),0)\big]\big[\dot{\ell}(\tau)-0\big]\,d\tau \geqslant 0 \\[4pt] \text{whenever } \rho_{\pi_t}(\ell, r) = (\ell, r).\end{aligned}\right\} \tag{3.10}$$

The integrand in (3.10) is of the form

$$\big[\phi_v(\ell,r_1) - \phi_v(\ell,r_2)\big]\big[r_1 - r_2\big],$$

and we may regard this as the change in force times the change in velocity at fixed length. If the function $r \mapsto \phi_v(\ell, r)$ is non-decreasing for each ℓ, then the change in force times the change in velocity (at fixed ℓ) is non-negative, and we obtain the result: *if ϕ_v is a non-decreasing function of its second argument, for fixed values of its first, then w_v has the dissipation property at every state.* In particular, if ϕ_v is of the form

$$\phi_v(\ell,r) = \mu(\ell)r \tag{3.11}$$

with $\mu(\ell) \geqslant 0$ for all ℓ in (ℓ_a, ℓ_b), then w_v has the dissipation property at every state.

We now turn to a study of necessary conditions in order that the Second Law (in the form (3.8)) hold. We imagine taking a process (π_t, ℓ) of a viscous filament and "retarding" it, i.e., making the rate of change of ℓ become smaller while the duration of the process becomes longer. This may be realized by replacing the function $\tau \mapsto \ell(\tau)$ on $[0, t]$ by the function $\tau \mapsto \ell_n(\tau) = \ell(\tau/n)$ on $[0, nt]$ (Figure 44a). In terms of process generators,

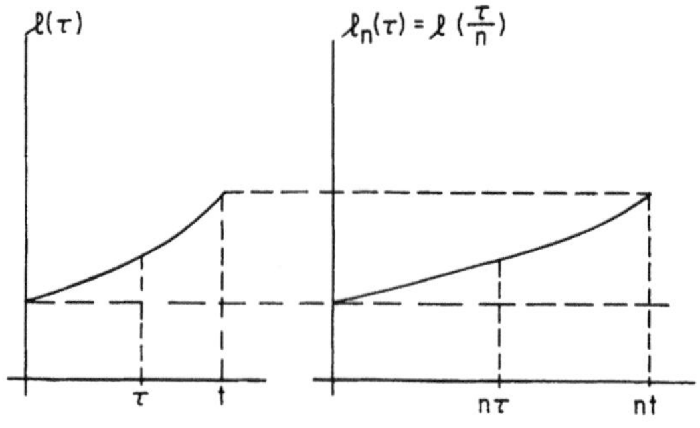

Figure 44a.

this amounts to defining π_{nt} by

$$\pi_{nt}(\tau) = \frac{\pi_t(\tau/n)}{n}, \qquad 0 \leqslant \tau < nt \tag{3.12}$$

and replacing $(\pi_t,(\ell,r))$ by $(\pi_{nt},(\ell,r))$. This procedure preserves the final length of the filament, because the first component of $(\rho_{\pi_{nt}}(\ell,r)$ is given by

$$\ell_n(nt) = \ell\left(\frac{nt}{n}\right) = \ell(t),$$

and $\ell(t)$ is the first component of $\rho_{\pi_t}(\ell,r)$. However, (3.12) shows that the second component of $\rho_{\pi_{nt}}(\ell,r)$ is $\pi_t(t-)/n$, whereas the second component of $\rho_{\pi_t}(\ell,r)$ is $\pi_t(t-)$, and this expresses the fact that the rates of change of length in the retarded processes $(\pi_{nt},(\ell,r))$ tend to zero as n tends to infinity. The next lemma gives important properties of w_v under retardation; in its proof we modify π_{nt} near $\tau = nt$ so that $(\pi_t,(\ell,r))$ and the modified retarded process have the *same* final states.

Lemma 3.1. *Let* $t, \ell_1,$ *and* ℓ_2 *be positive numbers and let* $\tau \mapsto \ell(\tau)$ *be a continuously differentiable function from* $[0,t]$ *into* Σ *such that* $\ell(0) = \ell_1$ *and* $\ell(t) = \ell_2$. *For every Helmholtz free energy function* Ψ *and for every* r_1, r_2 *in* $\mathbb{R}$, *there holds*

$$\Psi(\ell_2, r_2) - \Psi(\ell_1, r_1) = \int_0^t \phi_v(\ell(\tau), 0)\dot\ell(\tau)\, d\tau. \tag{3.13}$$

PROOF. For each positive integer n and r_2 in $\mathbb{R}$, we define $\tilde\ell_n: [0,nt] \to \mathbb{R}$ by

$$\tilde\ell_n(\tau) = \begin{cases} \ell\left(\dfrac{\tau}{n}\right), & 0 \leqslant \tau < nt - \dfrac{2}{n} \\[2mm] \ell\left(t - \dfrac{2}{n^2}\right) + n\left[\tau - \left(nt - \dfrac{2}{n}\right)\right]\left[\ell_2 - \dfrac{r_2}{n} - \ell\left(t - \dfrac{2}{n^2}\right)\right], & \\[2mm] & nt - \dfrac{2}{n} \leqslant \tau < nt - \dfrac{1}{n} \\[2mm] \ell_2 - \dfrac{r_2}{n} + n\left[\tau - \left(nt - \dfrac{1}{n}\right)\right]\left[\ell_2 - \left(\ell_2 - \dfrac{r_2}{n}\right)\right], & \\[2mm] & nt - \dfrac{1}{n} \leqslant \tau \leqslant nt \end{cases} \tag{3.14}$$

as shown in Figure 44b. For n sufficiently large, the values of $\tilde\ell_n$ lie in Σ, so the function $\tilde\pi_n$ on $[0,t)$ defined by

$$\tilde\pi_n(\tau) = \begin{cases} \dot{\tilde\ell}_n(\tau), & \tau \in [0,t), \tau \neq nt - \dfrac{2}{n}, nt - \dfrac{1}{n} \\[2mm] 0, & \tau = nt - \dfrac{2}{n}, \tau = nt - \dfrac{1}{n}, \end{cases}$$

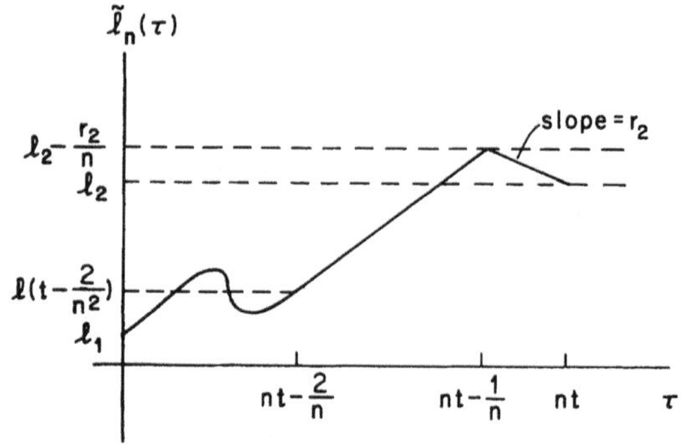

Figure 44b.

is a process generator, (ℓ_1, r_1) is in $\mathscr{D}(\tilde{\pi}_n)$ for every r_1 in $\mathbb{R}$, and $\rho_{\tilde{\pi}_n}(\ell_1, r_1) = (\ell_2, r_2)$. Therefore, for each Helmholtz free energy function Ψ, there holds

$$\Psi(\ell_2, r_2) - \Psi(\ell_1, r_1) \leqslant \omega_v(\tilde{\pi}_n, (\ell_1, r_1)),$$

i.e., for every n sufficiently large

$$\begin{aligned}\Psi(\ell_2, r_2) - \Psi(\ell_1, r_1) &\leqslant \int_0^{nt-(2/n)} \phi_v\left(\ell\left(\frac{\tau}{n}\right), \frac{1}{n}\dot{\ell}\left(\frac{\tau}{n}\right)\right)\frac{1}{n}\dot{\ell}\left(\frac{\tau}{n}\right) d\tau \\ &\quad + \int_{nt-(2/n)}^{nt} \phi_v(\tilde{\ell}_n(\tau), \dot{\tilde{\ell}}_n(\tau))\dot{\tilde{\ell}}_n(\tau)\, d\tau \\ &= I_1(n) + I_2(n). \end{aligned} \tag{3.15}$$

Consider now the relations:

$$\begin{aligned} I_1(n) &= \int_0^{(1/n)(nt-(2/n))} \phi_v\left(\ell(u), \frac{1}{n}\dot{\ell}(u)\right)\dot{\ell}(u)\, du \\ &= \int_0^t \phi_v\left(\ell(u), \frac{1}{n}\dot{\ell}(u)\right)\dot{\ell}(u)\, du \\ &\quad - \int_{t-(2/n^2)}^t \phi_v\left(\ell(u), \frac{1}{n}\dot{\ell}(u)\right)\dot{\ell}(u)\, du \\ &= \int_0^t \phi_v(\ell(u), 0)\dot{\ell}(u)\, du \\ &\quad + \int_0^t \left[\phi_v\left(\ell(u), \frac{1}{n}\dot{\ell}(u)\right) - \phi_v(\ell(u), 0)\right]\dot{\ell}(u)\, du \\ &\quad - \int_{t-(2/n^2)}^t \phi_v\left(\ell(u), \frac{1}{n}\dot{\ell}(u)\right)\dot{\ell}(u)\, du. \end{aligned} \tag{3.16}$$

The continuity of ϕ_v and boundedness of $\dot{\ell}$ imply that the last two integrals in (3.16) tend to zero as n tends to infinity, so that

$$\lim_{n\to\infty} I_1(n) = \int_0^t \phi_v(\ell(u), 0)\dot{\ell}(u)\, du. \tag{3.17}$$

To estimate $I_2(n)$, we note that

$$\ell_2 - \ell\left(t - \frac{2}{n^2}\right) = \ell(t) - \ell\left(t - \frac{2}{n^2}\right)$$
$$= \dot{\ell}(\xi)\frac{2}{n^2}$$

with $t - 2/n^2 < \xi < t$, and therefore

$$\dot{\tilde{\ell}}_n(\tau) = n\left(\dot{\ell}(\xi)\frac{2}{n^2} - \frac{r_2}{n}\right) = \frac{2}{n}\dot{\ell}(\xi) - r_2, \qquad nt - \frac{2}{n} \leqslant \tau \leqslant nt - \frac{1}{n}, \tag{3.18}$$

$$\dot{\tilde{\ell}}_n(\tau) = r_2, \qquad nt - \frac{1}{n} \leqslant \tau \leqslant nt. \tag{3.19}$$

The relations (3.18) and (3.19) imply that the integrand in $I_2(n)$ is bounded in absolute value, with an upper bound independent of n, so that

$$\lim_{n \to \infty} I_2(n) = 0;$$

this and (3.17) show that, for every r_1, r_2 in $\mathbb{R}$,

$$\Psi(\ell_2, r_2) - \Psi(\ell_1, r_1) \leqslant \int_0^t \phi_v(\ell(\tau), 0)\dot{\ell}(\tau)\,d\tau. \tag{3.20}$$

We can now apply (3.20) with $\tau \mapsto \ell(\tau)$ replaced by $\tau \mapsto \ell^r(\tau) = \ell(t - \tau)$ and with (ℓ_1, r_1) and (ℓ_2, r_2) interchanged. In fact, $\ell^r(0) = \ell_2$, $\ell^r(t) = \ell_1$ and r_1, r_2 can be chosen at will in (3.20), so that

$$\begin{aligned}\Psi(\ell_1, r_1) - \Psi(\ell_2, r_2) &\leqslant \int_0^t \phi_v(\ell^r(\tau), 0)\dot{\ell}^r(\tau)\,d\tau \\ &= \int_0^t \phi_v(\ell(t-\tau), 0)(-1)\dot{\ell}(t-\tau)\,d\tau \\ &= -\int_t^0 \phi_v(\ell(u), 0)\dot{\ell}(u)(-du) \\ &= -\int_0^t \phi_v(\ell(u), 0)\dot{\ell}(u)\,du,\end{aligned}$$

i.e.,

$$\Psi(\ell_2, r_2) - \Psi(\ell_1, r_1) \geqslant \int_0^t \phi_v(\ell(u), 0)\dot{\ell}(u)\,du. \tag{3.21}$$

By (3.20) and (3.21), relation (3.13) holds. □

We can write $\phi_{eq}(\ell)$ for $\phi_v(\ell, 0)$ and call ϕ_{eq} the *equilibrium response function* for the filament. Relation (3.13) then tells us that every Helmholtz free energy function Ψ is a potential for the action

$$(\pi_t, \ell) \mapsto \int_0^t \phi_{eq}(\ell(\tau))\dot{\ell}(\tau)\,d\tau. \tag{3.22}$$

Therefore, Ψ is independent of its second argument r and is an antiderivative for ϕ_{eq}. If Φ_{eq}° denotes the antiderivative of ϕ_{eq} which vanishes at ℓ_0, then we have the following theorem.

Theorem 3.1. *A viscous filament which satisfies the Second Law has exactly one Helmholtz free energy function* Ψ_{v}° *which vanishes at the state* $(\ell_0, 0)$. *It is given by*

$$\Psi_{\text{v}}^{\circ}(\ell, r) = \Phi_{\text{eq}}^{\circ}(\ell), \qquad (\ell, r) \in \Sigma \tag{3.23}$$

and, thus, is a potential for the action in (3.22) *and does not depend upon its second variable* r.

We now reexamine the statement that Ψ_{v}° is a lower potential for ω_{v}:

$$\Psi_{\text{v}}^{\circ}(\ell_2, r_2) - \Psi_{\text{v}}^{\circ}(\ell_1, r_1) \leqslant \int_0^t \phi_{\text{v}}(\ell(\tau), \dot{\ell}(\tau))\dot{\ell}(\tau)\, d\tau$$

whenever $\tau \mapsto \ell(\tau)$ describes a process $(\pi_t, (\ell_1, r_1))$ for which $\rho_{\pi_t}(\ell_1, r_1) = (\ell_2, r_2)$. This relation and (3.13), with $\Psi = \Psi_{\text{v}}^{\circ}$, then yield

$$\left.\begin{array}{l} \displaystyle\int_0^t \left[\phi_{\text{v}}(\ell(\tau), \dot{\ell}(\tau)) - \phi_{\text{eq}}(\ell(\tau))\right]\dot{\ell}(\tau)\, d\tau \geqslant 0 \\ \text{for } \textit{every} \text{ process } (\pi_t, (\ell, r)), \end{array}\right\} \tag{3.24}$$

and the Second Law implies that the inequality in (3.10) holds for *all* processes. We are in a position to give a necessary and sufficient condition on ϕ_{v} in order that the Second Law holds.

Theorem 3.2. *The Second Law is satisfied if and only if* ϕ_{v} *obeys the relation*

$$(\phi_{\text{v}}(\ell, r) - \phi_{\text{v}}(\ell, 0))r \geqslant 0 \tag{3.25}$$

for every $(\ell, r) \in \Sigma$.

Proof. Examination of (3.25) and the statement of the Second Law (3.10) shows that (3.25) implies the validity of the Second Law. Conversely, if (3.10) holds, then we have shown above that (3.24) holds. Given (ℓ, r) in Σ, there exists a $t_0 > 0$ such that, for each t in $(0, t_0)$, the function

$$\pi_t(\tau) = r, \qquad \tau \in [0, t)$$

is a process generator with (ℓ, r) in $\mathscr{D}(\pi_t)$. For the process $(\pi_t, (\ell, r))$ the function $\tau \mapsto \ell(\tau)$ is given by

$$\ell(\tau) = \ell + \tau r,$$

and (3.24) yields for *every* $t \in (0, t_0)$:

$$\frac{1}{t}\int_0^t \left[\phi_{\text{v}}(\ell + \tau r, r) - \phi_{\text{v}}(\ell + \tau r, 0)\right] r\, d\tau \geqslant 0.$$

Letting t tend to zero and using the fundamental theorem of Calculus we obtain

$$[\phi_v(\ell+\tau r, r) - \phi_v(\ell+\tau r, 0)]\, r|_{\tau=0} \geqslant 0,$$

which is (3.25). □

Relation (3.25) tells us that

$$\phi_v(\ell, r) r \geqslant \phi_{eq}(\ell) r$$

for every state (ℓ, r), and this inequality is the statement that *the rate at which the force* $f = \phi_v(\ell, r)$ *does work on the filament is no less than the rate at which the "equilibrium force"* $f_{eq} := \phi_{eq}(\ell)$ *does work.* Stated in another way, *the Second Law is equivalent to the assertion that the rate at which the "extra force"* $\phi_v(\ell, r) - \phi_{eq}(\ell)$ *does work on the filament is non-negative.* We can interpret the extra force to be due to the viscous resistance to motion, alone. When $\phi_v(\ell, r) = \mu(\ell) r$, the extra force is $\mu(\ell) r$ itself, because $\phi_{eq}(\ell) = \phi_v(\ell, 0) = \mu(\ell) 0 = 0$, and *the Second Law is satisfied if and only if* $\mu(\ell)$ *is non-negative for every* ℓ.

4. Elastic–Perfectly Plastic Filaments

The filaments we shall now discuss exhibit many properties of metallic solids undergoing large deformations. The two most striking features are (1) the constraint on the amount of force the filament can sustain and (2) the residual deformations which occur in the filament. Moreover, the evolution of the state of the filament is invariant under changes in time scale (like elastic filaments and unlike viscous filaments), and these features combine to give a simple example of a system whose state space contains many paths, but only relatively few of these paths correspond to processes of the system.

To specify an elastic–perfectly plastic filament, we give positive numbers ℓ_a, ℓ_b, β and f_y which satisfy the inequalities

$$\ell_a < \ell_b \quad \text{and} \quad \ell_a + \frac{f_y}{\beta} < \ell_b - \frac{f_y}{\beta} \tag{4.1}$$

with β called the *elastic modulus* and f_y the *yield force* of the filament. The states of the filament are pairs of the form (ℓ, f) in which ℓ represents the length and f the force in the filament, as usual. The state space Σ has the shape of a parallelogram (Figure 45) and is described as follows:

$$\Sigma = \left\{ (\ell, f) \,\middle|\, |f| \leqslant f_y,\ \ell_a + \frac{f_y}{\beta} < \ell - \frac{f}{\beta} < \ell_b - \frac{f_y}{\beta} \right\}. \tag{4.2}$$

For each state (ℓ, f), $\ell_R(\ell, f) := \ell - f/\beta$ lies in the interval $(\ell_a + f_y/\beta, \ell_b - f_y/\beta)$ and is called the *residual length* (or the permanent length) of the

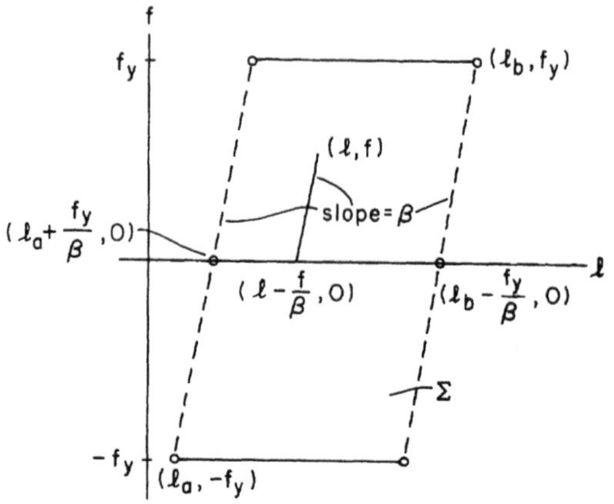

Figure 45.

filament for that state. It corresponds to the length obtained if the forces are removed from the filament in such a way that, at every instant, the change in force divided by change in length equals β. (A slight simplification occurs if the pairs (ℓ, ℓ_R) instead of (ℓ, f) are used to describe Σ, but we forego this in order to retain the variable f whose interpretation is more easily understood.)

To describe the set Π of process generators, we let $t > 0$ and π_t: $[0, t) \to \mathbb{R}$ be a piecewise continuous function which is also "piecewise of one sign," i.e., the interval $[0, t)$ can be partitioned into a finite number of subintervals (some possibly having zero length) on each of which π_t is everywhere positive, everywhere zero, or everywhere negative. For each function π_t of this type, we define the set $\mathscr{D}(\pi_t)$ to be the collection of all states (ℓ, f) for which the following initial value problem has a solution $\tau \mapsto (\ell(\tau), \mathrm{f}(\tau))$ on $[0, t]$ with values in Σ:

$$\text{(IVP)}\begin{cases} \dot{\ell}(\tau) = \pi_t(\tau) & (4.3)\\ \dot{\mathrm{f}}(\tau) = \begin{cases} 0, & \text{if } |\mathrm{f}(\tau)| = \mathrm{f}_y \text{ and } \mathrm{f}(\tau)\pi_t(\tau) \geqslant 0 \\ \beta\pi_t(\tau), & \text{otherwise,} \end{cases} & (4.4)\\ (\ell(0), \mathrm{f}(0)) = (\ell, \mathrm{f}). & (4.5) \end{cases}$$

Postponing for the moment a description of solutions of this problem, we define

$$\Pi = \{\pi_t\colon [0, t) \to \mathbb{R} \mid t > 0 \quad \text{and} \quad \mathscr{D}(\pi_t) \neq \varnothing\},$$

and note that, as for elastic and viscous filaments, the number $\pi_t(\tau)$ represents the rate of change of length $\dot{\ell}(\tau)$ of the filament in any process whose initial state is in $\mathscr{D}(\pi_t)$. (As usual, a dot near the top of a symbol denotes "derivative.")

By a *solution* of (IVP), we mean a continuous, piecewise continuously differentiable function which satisfies (4.3) and (4.4) at all but a finite set of points in $[0, t]$ and which satisfies the initial condition (4.5). The pair of equations

$$\left.\begin{aligned} \dot{\ell}(\tau) &= \pi_t(\tau) \\ \dot{f}(\tau) &= \beta\pi_t(\tau) \end{aligned}\right\} \tag{4.6}$$

is called the *elastic system* and the four relations

$$\left.\begin{aligned} \dot{\ell}(\tau) &= \pi_t(\tau), \qquad & f(\tau)\dot{\ell}(\tau) &\geqslant 0, \\ \dot{f}(\tau) &= 0, & |f(\tau)| &= f_y, \end{aligned}\right\} \tag{4.7}$$

are called the *plastic system*. Because π_t is piecewise of one sign, (IVP) can be studied by alternately solving initial value problems for the elastic system and for the plastic system. In particular, each such initial-value problem has a unique solution, and this implies that (IVP) has a unique solution. The solution trajectories of (4.6) are straight lines with slope β, while those of (4.7) are horizontal straight lines contained either in the locus $f = +f_y$ or in $f = -f_y$. Because of the condition $f(\tau)\pi_t(\tau) = f(\tau)\dot{\ell}(\tau) \geqslant 0$, $\ell(\tau)$ must increase on $f = +f_y$ and $\ell(\tau)$ must decrease on $f = -f_y$. A transition from a trajectory of (4.6) to one of (4.7) occurs when $f(\tau)$ reaches the value f_y (or $-f_y$) and $\dot{\ell}(\tau)$ is non-negative (or non-positive). A transition from a trajectory of (4.7) to one of (4.6) occurs when $f(\tau)$ is $+f_y$ (or $-f_y$) and $\dot{\ell}(\tau)$ becomes negative (or positive). Typical solution trajectories of (IVP) which lie in Σ are shown in Figure 46. Trajectories ① and ② contain two "elastic segments" and one "plastic segment" each, trajectory ③ contains one "elastic segment" and no "plastic segments," while ④ contains one "plastic segment" and no "elastic segments." Therefore, each element $(\pi_t, (\ell, f))$ of $\Pi \diamond \Sigma$ corresponds to a parameterized curve in Σ composed of segments of the type shown above, so that only special paths in Σ can be associated with processes of the system (Σ, Π). In particular, because trajectories on $f = +f_y$ move to the right and those on $f = -f_y$ move to the left, processes with plastic segments have no reversals. The trajectories illustrated above should make it obvious that the condition (S1) in the definition of a system with perfect accessibility is satisfied, and it is therefore easy to show that (Σ, Π) is a system with perfect accessibility. It is also clear that the state $(0, \ell_R(\ell, f))$ can be reached from (ℓ, f) via a process which contains one elastic segment and no plastic segments. In fact, $\ell_R(\ell(\tau), f(\tau))$ remains constant on elastic segments and changes on a plastic segment by amount equal to the change in $\ell(\tau)$ on that segment.

We can now write down a formula for the *work done on the filament* in a process $(\pi_t, (\ell, f))$,

$$\omega_{ep}(\pi_t, (\ell, f)) = \int_0^t f(\tau)\dot{\ell}(\tau)\, d\tau, \tag{4.8}$$

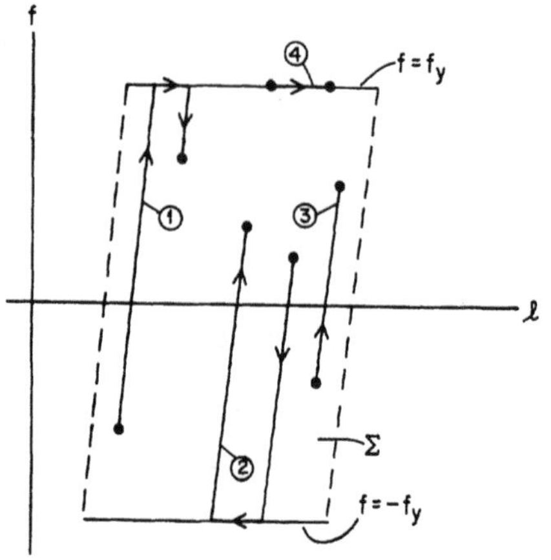

Figure 46.

where $\tau \mapsto (\ell(\tau), \mathrm{f}(\tau))$ is the solution of (IVP) for the given pair $(\pi_t, (\ell, \mathrm{f}))$, and it requires only a routine calculation to verify that ω_{ep} *is an action for* (Σ, Π). If $\Gamma(\pi_t, (\ell, \mathrm{f}))$ denotes the oriented piecewise smooth curve in Σ parameterized by $\tau \mapsto (\ell(\tau), \mathrm{f}(\tau))$, then we can express the work done as a line integral

$$\omega_{\mathrm{ep}}(\pi_t, (\ell, \mathrm{f})) = \int_{\Gamma(\pi_t, (\ell, \mathrm{f}))} \bar{\mathrm{f}}\, d\bar{\ell}. \tag{4.9}$$

Of course, $\Gamma(\pi_t, (\ell, \mathrm{f}))$ is a solution trajectory of (IVP) lying in Σ, so it must consist entirely of elastic and plastic segments. In the special case where $(\Gamma(\pi_t, (\ell, \mathrm{f}))$ is a simple closed curve, then that curve must be negatively oriented, and Green's Theorem yields

$$\omega_{\mathrm{ep}}(\pi_t, (\ell, \mathrm{f})) = \oint_{\Gamma(\pi_t, (\ell, \mathrm{f}))} \bar{\mathrm{f}}\, d\bar{\ell} = \int_{\mathrm{Int}\, \Gamma(\pi_t, (\ell, \mathrm{f}))} d\bar{\ell}\, d\bar{\mathrm{f}},$$

so that $\omega_{\mathrm{ep}}(\pi_t, (\ell, \mathrm{f}))$ *is the area of the interior of* $\Gamma(\pi_t, (\ell, \mathrm{f}))$ *when the process* $(\pi_t, (\ell, \mathrm{f}))$ *determines a simple closed curve*. Therefore, ω_{ep} is non-negative on cycles of this type, and this fact leads us to conjecture that ω_{ep} is non-negative on every cycle.

Theorem 4.1. *Every elastic-plastic filament obeys the Second Law. In fact, for every process* $(\pi_t, (\ell, \mathrm{f}))$ *there holds*

$$\omega_{\mathrm{ep}}(\pi_t, (\ell, \mathrm{f})) \geqslant \frac{\mathrm{f}(t)^2}{2\beta} - \frac{\mathrm{f}^2}{2\beta}, \tag{4.10}$$

where $\mathrm{f}(t)$ *is the force on the filament at the end of the process* $(\pi_t, (\ell, \mathrm{f}))$.

Of course, (4.10) implies that ω_{ep} has the dissipation property, because the right-hand side of (4.10) vanishes when the process is a cycle.

PROOF OF (4.10). Each process $(\pi_t,(\ell,\mathrm{f}))$ induces a partition of $[0,t)$ into adjacent intervals of $\mathrm{E}_1, P_1, \mathrm{E}_2, \ldots,$ finite in number, such that the elastic system (4.6) holds on the intervals labeled with E and the plastic system (4.7) holds on the intervals labeled with P. Therefore, we can write (4.8) in the form

$$\omega_{\mathrm{ep}}(\pi_t,(\ell,\mathrm{f})) = \sum_{\mathrm{E}_k} \int_{\mathrm{E}_k} \mathrm{f}(\tau)\frac{\dot{\mathrm{f}}(\tau)}{\beta}\,d\tau + \sum_{P_j}\int_{P_j} \mathrm{f}(\tau)\dot{\ell}(\tau)\,d\tau$$

$$\geqslant \sum_{\mathrm{E}_k} \left.\frac{\mathrm{f}(\tau)^2}{2\beta}\right|_{a_k}^{b_k}, \tag{4.11}$$

where we have used (4.6) and (4.7) and have written $\mathrm{E}_k = [a_k, b_k]$ in the last step. Consider now intervals E_k and E_{k+1}; they are separated by one of the intervals P_j, so that $P_j = [b_k, a_{k+1}]$ and

$$\mathrm{f}^2(b_k) = \mathrm{f}^2(a_{k+1}) = \mathrm{f}_y^2.$$

Therefore, the last sum in (4.11) is telescoping and (4.10) follows immediately when both the first and last intervals in the partition are of elastic type. Some reflection shows that (4.10) follows from (4.11) even when one or both of the first and last intervals is of plastic type. □

Thus, not only elastic filaments, but also elastic-plastic filaments have the property that the Second Law holds "automatically," i.e., as a consequence of the properties which define these filaments, so that the positive numbers ℓ_a, ℓ_b, β and f_y are not subject to any restrictions arising from that law. It remains only to study the class $\mathscr{L}^\circ$ of Helmholtz free energy functions which vanish at a preassigned "standard" state $(\ell_0, 0)$. Unlike elastic and viscous filaments, elastic-plastic filaments do *not* have the property that $\mathscr{L}^\circ$ is a singleton. The rest of this section will be devoted to verifying this assertion and to describing precisely how large the set $\mathscr{L}^\circ$ can be.

Theorem 4.1 permits us to apply Theorem 4.1 of Chapter V to conclude that $\mathscr{L}^\circ$ is non-empty. Moreover, the trivial process generator $\bar{\pi}_t$, defined by

$$\bar{\pi}_t(\tau) = 0, \qquad \tau \in [0,t),$$

leaves every state unaltered and causes no work to be done, so we may also apply the second part of Theorem 4.1 to obtain the smallest and largest elements of $\mathscr{L}^\circ$:

$$\hat{\Psi}^\circ(\ell,\mathrm{f}) = -\inf\{\omega_{\mathrm{ep}}(\pi_t,(\ell,\mathrm{f})) \mid \rho_{\pi_t}(\ell,\mathrm{f}) = (\ell_0,0)\}, \tag{4.12}$$

$$\Psi^\circ(\ell,\mathrm{f}) = \inf\{\omega_{\mathrm{ep}}(\pi_t,(\ell_0,0)) \mid \rho_{\pi_t}(\ell_0,0) = (\ell,\mathrm{f})\}. \tag{4.13}$$

Our first task will be to obtain more explicit formulas for these functions. For example, suppose that we wish to evaluate the infimum in (4.13). We employ the first line of (4.11), with $\ell = \ell_0$ and $\mathrm{f} = 0$, and use (4.6) and (4.7) to obtain the formula

$$\omega_{\mathrm{ep}}(\pi_t,(\ell_0,0)) = \sum_{\mathrm{E}_k} \frac{\mathrm{f}(\tau)^2}{2\beta}\bigg|_{a_k}^{b_k} + \sum_{P_j}\int_{P_j} |\mathrm{f}(\tau)||\dot{\ell}(\tau)|\,d\tau$$

$$= \frac{\mathrm{f}(t)^2}{2\beta} + \mathrm{f}_y \sum_{\mathrm{P}_j}\int_{P_j} |\dot{\ell}(\tau)|\,d\tau. \tag{4.14}$$

When $\rho_{\pi_t}(\ell_0,0)=(\ell,\mathrm{f})$, then $\mathrm{f}(t)=\mathrm{f}$; moreover, each term in the sum over P_j represents the length of the segment on $|\mathrm{f}|=\mathrm{f}_y$ traversed during P_j. Therefore, (4.14) yields the implication

$$\rho_{\pi_t}(\ell_0,0)=(\ell,\mathrm{f}) \Rightarrow \omega_{\mathrm{ep}}(\pi_t,(\ell_0,0)) = \frac{\mathrm{f}^2}{2\beta} + \mathrm{f}_y \mathrm{L}_p(\pi_t,(\ell_0,0)) \tag{4.15}$$

where $\mathrm{L}_p(\pi_t,(\ell_0,0))$ is the sum of the lengths of the plastic segments of $\Gamma(\pi_t,(\ell_0,0))$. To obtain the infimum in (4.13), we must minimize $\mathrm{L}_p(\pi_t,(\ell_0,0))$ subject to the constraint $\rho_{\pi_t}(\ell_0,0)=(\ell,\mathrm{f})$. It is clear from the Figure 47 that the minimum value of $\mathrm{L}_p(\pi_t,(\ell_0,0))$ is the absolute value of the difference in residual lengths of the initial and final states of the process $(\pi_t,(\ell_0,0))$, and this implies

$$\Psi^{\circ}(\ell,\mathrm{f}) = \frac{\mathrm{f}^2}{2\beta} + \mathrm{f}_y\left|\ell - \frac{\mathrm{f}}{\beta} - \ell_0\right|. \tag{4.16}$$

A similar argument shows that

$$\hat{\Psi}^{\circ}(\ell,\mathrm{f}) = \frac{\mathrm{f}^2}{2\beta} - \mathrm{f}_y\left|\ell - \frac{\mathrm{f}}{\beta} - \ell_0\right|, \tag{4.17}$$

and we have proven that the *largest and smallest elements of the convex set* $\mathscr{L}^{\circ}$ *are given by the formulas* (4.16) *and* (4.17).

We observe from (4.10) that $\overline{\Psi}$ given by

$$\overline{\Psi}(\ell,\mathrm{f}) := \frac{\mathrm{f}^2}{2\beta} = \frac{\Psi^{\circ}(\ell,\mathrm{f}) + \hat{\Psi}^{\circ}(\ell,\mathrm{f})}{2} \tag{4.18}$$

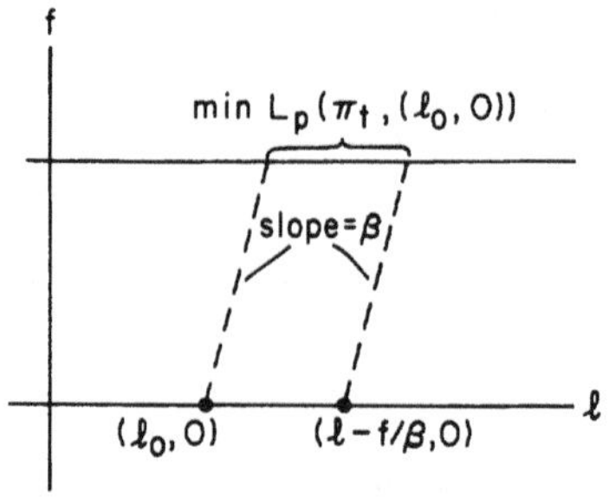

Figure 47.

is an element of $\mathscr{L}^\circ$, and (4.16)–(4.18) suggest that every Helmholtz free energy function contains the term $f^2/2\beta$. To study this conjecture, let Ψ be in $\mathscr{L}^\circ$, and let (ℓ_1, f_1) and (ℓ_2, f_2) be states with the same residual deformation, i.e.,

$$\ell_1 - \frac{f_1}{\beta} = \ell_R(\ell_1, f_1) = \ell_R(\ell_2, f_2) = \ell_2 - \frac{f_2}{\beta}. \tag{4.19}$$

If we restrict attention to processes $(\pi_t, (\ell_1, f_1))$, with $\rho_{\pi_t}(\ell_1, f_1) = (\ell_2, f_2)$, which have no plastic segments, than our analysis of elastic filaments tells us that

$$\Psi(\ell_2, f_2) - \Psi(\ell_1, f_1) = w_{ep}(\pi_t, (\ell_1, f_1)), \tag{4.20}$$

and the calculation which gave us (4.11) (where now there are no intervals P_j and one interval E_k) yields

$$w_{ep}(\pi_t, (\ell_1, f_1)) = \frac{f_2^2}{2\beta} - \frac{f_1^2}{2\beta}. \tag{4.21}$$

Now, let (ℓ, f) be any state and note that

$$\ell_R(\ell, f) = \ell_R\left(\ell - \frac{f}{\beta}, 0\right) = \ell - \frac{f}{\beta},$$

so that we may apply (4.20) and (4.21), with $(\ell_1, f_1) = \left(\ell - \frac{f}{\beta}, 0\right)$ and $(\ell_2, f_2) = (\ell, f)$, to obtain

$$\Psi(\ell, f) - \Psi\left(\ell - \frac{f}{\beta}, 0\right) = \frac{f^2}{2\beta} - \frac{0^2}{\beta},$$

i.e.,

$$\Psi(\ell, f) = \frac{f^2}{2\beta} + \Psi(\ell_R(\ell, f), 0), \tag{4.22}$$

and (4.22) shows that *every element Ψ of $\mathscr{L}^\circ$ is of the form*

$$\Psi(\ell, f) = \frac{f^2}{2\beta} + \Psi_R(\ell_R(\ell, f)), \tag{4.23}$$

where the function $\ell_R \mapsto \Psi_R(\ell_R)$ has domain the open interval

$$\ell_a + \frac{f_y}{\beta} < \ell_R < \ell_b - \frac{f_y}{\beta} \tag{4.24}$$

and vanishes when ℓ_R equals ℓ_0. The functions $\Psi_R^\circ, \hat{\Psi}_R^\circ$ and $\overline{\Psi}_R$ in the formulas (4.16)–(4.18) are given by

$$\Psi_R^\circ(\ell_R) = f_y|\ell_R - \ell_0|, \tag{4.25}$$

$$\hat{\Psi}_R^\circ(\ell_R) = -f_y|\ell_R - \ell_0|, \tag{4.26}$$

$$\overline{\Psi}_R(\ell_R) = 0, \tag{4.27}$$

respectively. The graphs of these functions are shown in Figure 48.

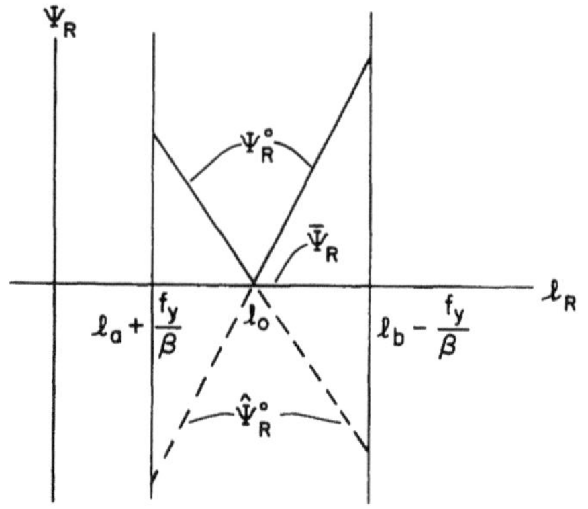

Figure 48.

The discussion above has reduced the problem of delimiting the class $\mathscr{L}^{\circ}$ to one of describing the class $\mathscr{L}_{\mathrm{R}}^{\circ}$ of functions Ψ_{R} which make the function Ψ defined in (4.23) a Helmholtz free energy function which vanishes at $(\ell_0, 0)$. At this point, we can assert that $\Psi_{\mathrm{R}}^{\circ}, \hat{\Psi}_{\mathrm{R}}^{\circ}$, and $\overline{\Psi}_{\mathrm{R}}$ are functions in $\mathscr{L}_{\mathrm{R}}^{\circ}$; in fact, $\Psi_{\mathrm{R}}^{\circ}$ and $\hat{\Psi}_{\mathrm{R}}^{\circ}$ are the largest and smallest elements of $\mathscr{L}_{\mathrm{R}}^{\circ}$. The next result gives the information required to characterize functions in $\mathscr{L}_{\mathrm{R}}^{\circ}$.

Theorem 4.2. *A function* $\Psi_{\mathrm{R}}\colon \left(\ell_a + \dfrac{\mathrm{f}_y}{\beta}, \ell_b - \dfrac{\mathrm{f}_y}{\beta}\right) \to \mathbb{R}$ *is in* $\mathscr{L}_{\mathrm{R}}^{\circ}$ *if and only if it vanishes at* ℓ_0 *and is Lipschitz continuous with Lipschitz constant* f_y. *In other words,* Ψ_{R} *is in* $\mathscr{L}_{\mathrm{R}}^{\circ}$ *if and only if there hold*

(i) $\Psi_{\mathrm{R}}(\ell_0) = 0$

(ii) $|\Psi_{\mathrm{R}}(\ell_{\mathrm{R}}'') - \Psi_{\mathrm{R}}(\ell_{\mathrm{R}}')| \leqslant \mathrm{f}_y |\ell_{\mathrm{R}}'' - \ell_{\mathrm{R}}'|$ *for all* $\ell_{\mathrm{R}}', \ell_{\mathrm{R}}''$ *in the interval* $\left(\ell_a + \dfrac{\mathrm{f}_y}{\beta}, \ell_b - \dfrac{\mathrm{f}_y}{\beta}\right)$.

PROOF. First we suppose that Ψ_{R} is in $\mathscr{L}_{\mathrm{R}}^{\circ}$ and let $\ell_{\mathrm{R}}' < \ell_{\mathrm{R}}''$ be in $\left(\ell_a + \dfrac{\mathrm{f}_y}{\beta}, \ell_b - \dfrac{\mathrm{f}_y}{\beta}\right)$. The function Ψ in (4.23) is then a Helmholtz free energy function, so that

$$\begin{aligned}\Psi_{\mathrm{R}}(\ell_{\mathrm{R}}'') - \Psi_{\mathrm{R}}(\ell_{\mathrm{R}}') &= \Psi(\ell_{\mathrm{R}}'', 0) - \Psi(\ell_{\mathrm{R}}', 0)\\ &\leqslant \omega_{\mathrm{ep}}(\pi_t, (\ell_{\mathrm{R}}', 0))\end{aligned} \tag{4.28}$$

for every process $(\pi_t, (\ell_{\mathrm{R}}', 0))$ such that $\rho_{\pi_t}(\ell_{\mathrm{R}}', 0) = (\ell_{\mathrm{R}}'', 0)$. If we choose π_t so that $\Gamma(\pi_t, (\ell_{\mathrm{R}}', 0))$ is the path shown in Figure 49, then the first relation in (4.11) yields

$$\omega_{\mathrm{ep}}(\pi_t, (\ell_{\mathrm{R}}', 0)) = \mathrm{f}_y(\ell_{\mathrm{R}}'' - \ell_{\mathrm{R}}'),$$

and (4.28) tells us that

$$\Psi_{\mathrm{R}}(\ell_{\mathrm{R}}'') - \Psi_{\mathrm{R}}(\ell_{\mathrm{R}}') \leqslant \mathrm{f}_y(\ell_{\mathrm{R}}'' - \ell_{\mathrm{R}}') = \mathrm{f}_y|\ell_{\mathrm{R}}'' - \ell_{\mathrm{R}}'|. \tag{4.29}$$

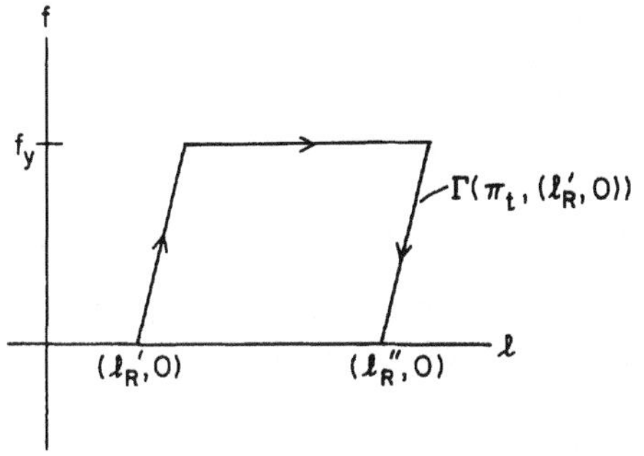

Figure 49.

If we repeat the same argument for a process generator $\tilde{\pi}_t$ such that $\Gamma(\tilde{\pi}_t, (\ell''_R, 0))$ is the path in Figure 50 then we conclude that

$$\Psi_R(\ell'_R) - \Psi_R(\ell''_R) \leqslant w_{ep}(\tilde{\pi}_t, (\ell''_R, 0))$$
$$= (-f_y)(\ell'_R - \ell''_R) = f_y|\ell''_R - \ell'_R|, \qquad (4.30)$$

and (4.29) together with (4.30) yield

$$|\Psi_R(\ell''_R) - \Psi_R(\ell'_R)| \leqslant f_y|\ell''_R - \ell'_R|.$$

Hence, every element of $\mathscr{L}^\circ_R$ obeys (ii) and, by definition of $\mathscr{L}^\circ_R$, (i).

Conversely, suppose that a function Ψ_R obeys (i) and (ii) and consider the function Ψ in (4.23). Clearly, $\Psi(\ell_0, 0) = 0 + \Psi_R(\ell_R(\ell_0, 0)) = \Psi_R(\ell_0) = 0$. For any two states (ℓ', f') and (ℓ'', f'') and for every choice of process generator π_t such that $\rho_{\pi_t}(\ell', f') = (\ell'', f'')$, a calculation similar to those given in (4.11), (4.14), and (4.15) yields

$$w_{ep}(\pi_t, (\ell', f')) = \frac{f''^2}{2\beta} - \frac{f'^2}{2\beta} + f_y \sum_{P_j} \int_{P_j} |\dot{\ell}(\tau)|\, d\tau$$
$$\geqslant \frac{f''^2}{2\beta} - \frac{f'^2}{2\beta} + f_y|\ell_R(\ell'', f'') - \ell_R(\ell', f')|. \qquad (4.31)$$

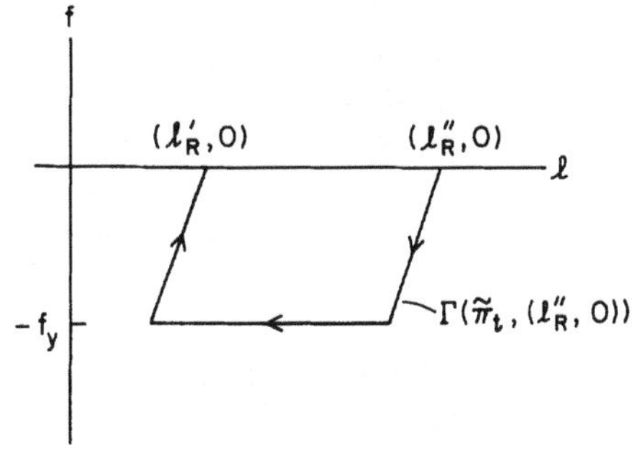

Figure 50.

According to (ii), there holds

$$f_y|\ell_R(\ell'',f'')-\ell_R(\ell',f')| \geqslant \Psi_R(\ell_R(\ell'',f''))-\Psi_R(\ell_R(\ell',f')), \quad (4.32)$$

so that (4.23), (4.31), and (4.32) imply

$$\omega_{ep}(\pi_t,(\ell',f')) \geqslant \Psi(\ell'',f'')-\Psi(\ell',f').$$

Therefore, Ψ is a Helmholtz free energy function which vanishes at $(\ell_0,0)$, and this completes the proof of Theorem 4.2. □

To summarize, Theorem 4.2 tells us that *every Helmholtz free energy function which vanishes at* $(\ell_0,0)$ *is of the form*

$$\Psi(\ell,f)=\frac{f^2}{2\beta}+\Psi_R\left(\ell-\frac{f}{\beta}\right), \quad (4.33)$$

where the function Ψ_R *obeys* (i) *and* (ii) *in the statement of Theorem 4.2*. Consequently, the set $\mathscr{L}^\circ$ contains elements which are *not* convex combinations $\alpha\Psi^\circ+(1-\alpha)\hat{\Psi}^\circ$ of the largest and smallest elements Ψ° and $\hat{\Psi}^\circ$. An example of such an element would be the function given by

$$\Psi(\ell,f)=\frac{f^2}{2\beta}+\frac{f_y^2}{2\beta}\sin\left(\frac{2\beta}{f_y}\left(\ell-\frac{f}{\beta}-\ell_0\right)\right). \quad (4.34)$$

5. Phase Transitions in Elastic and Viscous Filaments

We consider now an elastic filament, as defined in Section 2 of this chapter, and we specify not only the function ϕ_e, which gives the force f in the filament as a function of its length ℓ, but also an action $(\pi_t,\ell)\mapsto h_e(\pi_t,\ell)$ which assigns to each process (π_t,ℓ) the net heat gained by the filament in that process.

We say that the filament *admits a phase transition* at a value of force f_{tr} over an interval $[\ell_1,\ell_2]$ if the function ϕ_e has the form displayed in Figure 51. The horizontal portion of the graph of ϕ_e illustrates the capability of the filament of undergoing a process in which its length changes and, not only its temperature, but also the applied force is constant. The restrictions of ϕ_e to $(\ell_a,\ell_1]$ and to $[\ell_2,\ell_b)$ define two elastic filaments which we call distinct *phases* (phase 1 and phase 2) of the given filament. A process which starts at ℓ_1 and ends at ℓ_2, or *vice versa*, is said to cause a *change of phase* in the filament.

In order to justify the nomenclature "homogeneous" for a filament which undergoes a change of phase, we must imagine that the heat absorbed or emitted is distributed uniformly over the length of the filament. If the filament thereby attains a state ℓ between ℓ_1 and ℓ_2, we may regard each point along its length as being occupied by a mixture of the two phases,

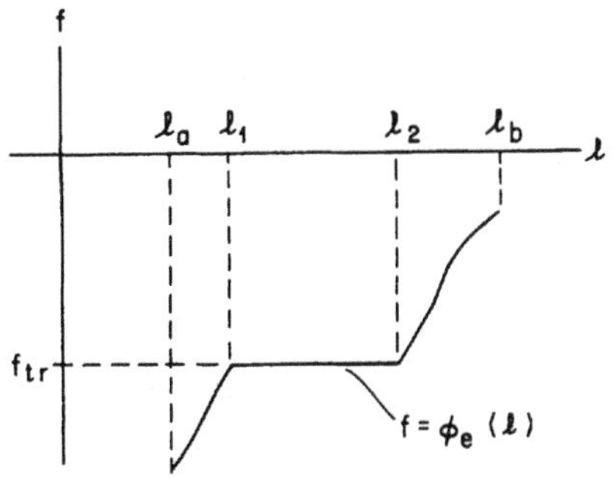

Figure 51.

with the composition of the mixture a constant along the filament. Even in cases where the assumption of homogeneity is not appropriate, the analysis we give may still be applicable, because it depends only on the form of the function ϕ_e and other assumptions on the heat supply, and these assumptions are valid in cases where the filament consists of two homogeneous parts, one corresponding to each phase.

We invoke now the First Law, taking it to be the following assertion:

$$\rho_{\pi_t}\ell = \ell \Rightarrow \omega_e(\pi_t, \ell) + h_e(\pi_t, \ell) = 0. \tag{5.1}$$

(Here we choose units so that Joule's constant J equals one.) If we recall the result on elastic filaments proven earlier, we can immediately restate (5.1) in the form:

$$\rho_{\pi_t}\ell = \ell \Rightarrow h_e(\pi_t, \ell) = 0. \tag{5.2}$$

In fact, we showed that ω_e vanishes on cycles, so (5.1) and (5.2) are equivalent statements. We note that for isothermal cycles of a homogeneous fluid body, this property of the heat gained follows from its definition, without appeal to the First Law. Here, (5.2) is a consequence of the formula (2.3) for ω_e and the First Law. From the First Law we can at once conclude that there is a function Λ_e: $(\ell_a, \ell_b) \to \mathbb{R}$ such that

$$\rho_{\pi_t}\ell' = \ell'' \Rightarrow h_e(\pi_t, \ell') = \Lambda_e(\ell'') - \Lambda_e(\ell'). \tag{5.3}$$

We write λ_{12} for the number $\Lambda_e(\ell_2) - \Lambda_e(\ell_1)$ and call it the *latent heat* for the change of phase from ℓ_1 to ℓ_2. According to (5.3), the latent heat can be produced by performing any process starting at ℓ_1 and ending at ℓ_2. (Of course, all processes under consideration already are assumed to be isothermal.) If E is an internal energy function for the filament, then the First and Second Laws yield the formulae

$$\begin{aligned}\lambda_{12} &= E(\ell_2) - E(\ell_1) - \omega_e(\pi_t, \ell_1)\\ &= (E(\ell_2) - E(\ell_1)) - (\Psi(\ell_2) - \Psi(\ell_1))\\ &= (E(\ell_2) - E(\ell_1)) - f_{tr}(\ell_2 - \ell_1)\\ &= (E(\ell) - \phi_e(\ell)\ell)\big|_{\ell_1}^{\ell_2}.\end{aligned} \tag{5.4}$$

(The function $\ell \mapsto \mathrm{E}(\ell)-\phi_{\mathrm{e}}(\ell)\ell$ corresponds to the function $(V,\theta)\mapsto \mathrm{E}(V,\theta)+p(V,\theta)V$, called the *enthalpy* in classical thermodynamics.) The formula $(5.4)_3$ can be written as a formula for $\mathrm{E}(\ell_2)$ in terms of $\mathrm{E}(\ell_1)$ and numbers which describe the phase transition:

$$\mathrm{E}(\ell_2)=\mathrm{E}(\ell_1)+\lambda_{12}+\mathrm{f}_{\mathrm{tr}}(\ell_2-\ell_1). \tag{5.5}$$

We regard (5.5) as a means by which we may "continue" an internal energy function for phase 1 into one for phase 2, for this relation specifies the arbitrary constant for the second internal energy function in terms of that for the first.

We turn now to a discussion of phase transitions for a viscous filament. Here, we specify not only the function $(\ell,r)\mapsto\phi_{\mathrm{v}}(\ell,r)$, which gives the force $\mathrm{f}=\phi_{\mathrm{v}}(\ell,r)$ required to have the filament at length ℓ when its length changes at a rate r, but also an action $(\pi_t,(\ell,r))\mapsto \hbar_{\mathrm{v}}(\pi_t,(\ell,r))$ which gives the heat gained in processes of the filament. As we shall see below, there is no meaningful concept of latent heat for changes of state of a viscous filament, simply because the action $\hbar_{\mathrm{v}}$ need not vanish on cycles, as was the case for $\hbar_{\mathrm{e}}$. However, there is a natural way of obtaining an elastic filament from a viscous one by considering only states of the form $(\ell,0)$. On such states, the function ϕ_{v} reduces to the function ϕ_{eq} defined by

$$\phi_{\mathrm{eq}}(\ell)=\phi_{\mathrm{v}}(\ell,0), \tag{5.6}$$

and w_{v} is replaced by the action w_{eq} defined by (3.22). Moreover, the proof of the Lemma 3.1 on viscous filaments shows that the value of w_{v} on a process tends to the value of w_{eq} on that process as the process is retarded indefinitely. This permits us to say that the elastic filament approximates the viscous filament in slow processes, and we call the elastic filament determined by $\phi_{\mathrm{eq}}\colon(\ell_a,\ell_b)\to\mathbb{R}$ in (5.6) the *approximating elastic filament* associated with the given viscous filament.

Our plan here is to assume that the approximating elastic filament admits a phase transition between states ℓ_1 and ℓ_2, in the sense just described, and to compare the heat gained in processes of the viscous filament, starting at $(\ell_1,0)$ and ending at $(\ell_2,0)$, with the latent heat λ_{12} for the change of phase from ℓ_1 to ℓ_2. In order to carry out this plan, it is necessary to assume that, for each process $(\pi_t,(\ell,r))$, the heat gained $\hbar_{\mathrm{v}}(\pi_t,(\ell,r))$ has a limit under retardations. For many natural choices of $\hbar_{\mathrm{v}}$, this can be verified and the limit turns out to depend only on the path $\Gamma(\pi_t,\ell)$ determined by the process (π_t,ℓ) associated with the approximating elastic filament. Therefore, *we assume that, for each process* $(\pi_t,(\ell,r))$, *the sequence* $\mathrm{n}\mapsto\hbar_{\mathrm{v}}(\tilde{\pi}_{\mathrm{n}},(\ell,r))$, *with* $\tilde{\pi}_{\mathrm{n}}$ *defined below* (3.14), *converges to a number* $\hbar_{\mathrm{eq}}(\Gamma(\pi_t,\ell))$.

We take the First Law to be the assertion

$$\rho_{\pi_t}(\ell',r')=(\ell',r')\Rightarrow w_{\mathrm{v}}(\pi_t,(\ell',r'))+\hbar_{\mathrm{v}}(\pi_t,(\ell',r'))=0, \tag{5.7}$$

and this is equivalent to the existence of an internal energy function E_{v}:

$(\ell_a, \ell_b)\times\mathbb{R} \to \mathbb{R}$, i.e., a function obeying the relation

$$E_v(\ell'', r'') - E_v(\ell', r') = \omega_v(\pi_t, (\ell', r')) + h_v(\pi_t, (\ell', r')) \qquad (5.8)$$

whenever $\rho_{\pi_t}(\ell', r') = (\ell'', r'')$. Let us apply (5.8) to a retardation $(\tilde{\pi}_n, (\ell', r'))$ of $(\pi_t, (\ell', r'))$. The left-hand side of (5.8) is independent of n, because $\rho_{\tilde{\pi}_n}(\ell', r') = (\ell'', r'')$ for all n. As n tends to infinity the right-hand side tends to

$$\int_0^t \phi_{eq}(\ell(\tau))\dot{\ell}(\tau)\, d\tau + h_{eq}(\Gamma(\pi_t, \ell')),$$

where $\ell(\tau) = \ell' + \int_0^\tau \pi_t(\xi)\, d\xi$, and this sum depends only on $\Gamma(\pi_t, \ell')$. This yields the next result.

Theorem 5.1. *Every internal energy function* E_v *for a viscous filament satisfies the relation*

$$\mathrm{E}_v(\ell'', r'') - \mathrm{E}_v(\ell', r') = \int_0^t \phi_{eq}(\ell(\tau))\dot{\ell}(\tau)\, d\tau + h_{eq}(\Gamma(\pi_t, \ell')) \qquad (5.9)$$

for each process $(\pi_t, (\ell', r'))$ *which ends at* (ℓ'', r''). *In particular,* E_v *is independent of the second variable.*

We now apply both (5.8) and (5.9) in the case where $\ell' = \ell_1$, $\ell'' = \ell_2$, $r' = r'' = 0$ and make use of (3.7) and (5.6) to obtain the equation

$$h_v(\pi_t, (\ell_1, 0)) = h_{eq}(\Gamma(\pi_t, \ell_1)) - \int_0^t [\phi_v(\ell(\tau), \dot{\ell}(\tau-)) - \phi_v(\ell(\tau), 0)]\dot{\ell}(\tau-)\, d\tau. \qquad (5.10)$$

Because $\Gamma(\pi_t, \ell_1)$ starts at ℓ_1 and ends at ℓ_2, we can identify $h_{eq}(\Gamma(\pi_t, \ell_1))$ as the latent heat λ_{12} for the approximating elastic filament. Moreover, Theorem 3.2 tells us that the integral in (5.10) is non-negative, so we obtain from (5.10) the inequality

$$h_v(\pi_t, (\ell_1, 0)) \leqslant \lambda_{12}, \qquad (5.11)$$

whenever $\rho_{\pi_t}(\ell_1, 0) = (\ell_2, 0)$.

We suppose now that λ_{12} is positive, as would be the case for a solid-to-liquid or a liquid-to-vapor change of phase. In this case, (5.11) says that *a viscous filament always absorbs less heat when transformed from* $(\ell_1, 0)$ *to* $(\ell_2, 0)$ *than does the approximating elastic filament for the change of phase from* ℓ_1 *to* ℓ_2. Therefore, we may think of the viscous forces as aiding the change of phase of the approximating elastic filament, as would be the case if one gently stirs a solid-liquid mixture in order to hasten the melting of the solid. A similar analysis applies to a change of phase from ℓ_2 to ℓ_1, which corresponds to a liquid-to-solid or vapor-to-liquid change of phase. In this case the corresponding latent heat λ_{21} equals $-\lambda_{12}$ and so is negative, (5.11) is replaced by

$$\mathscr{h}_{v}(\pi_t,(\ell_2,0)) \leqslant \lambda_{21} < 0$$

whenever $\rho_{\pi_t}(\ell_2,0) = (\ell_1,0)$ and, therefore, *more heat is emitted when a viscous filament is transformed from* $(\ell_2,0)$ *to* $(\ell_1,0)$ *than for the change of phase from* ℓ_2 *to* ℓ_1 *of the approximating elastic filament.*

CHAPTER VII

Homogeneous Fluid Bodies with Viscosity

1. Bodies with Viscosity as Thermodynamical Systems

In Chapter I we studied an important class of physical systems, homogeneous fluid bodies, employing concepts and methods already available to the pioneers of thermodynamics. The systems to be discussed here, called *homogeneous fluid bodies with viscosity*, are intended to provide, first of all, a more complete model of liquids and gases than do homogeneous fluid bodies and, second of all, a refinement of the notion of homogeneous viscous filaments introduced in Chapter VI. In contrast to the approach taken in Chapter I, we here shall avail ourselves of the modern conceptual framework developed in Chapters II through V. This will provide not only a review of that material, as it is applied to a specific class of physical systems, but also a useful comparison of the classical and modern approaches.

Our goal in this section is to use Definition 1.1 of Chapter III and Definition 4.1 of Chapter IV as a framework for describing a homogeneous fluid body with viscosity. We recall that each of these definitions requires that we prescribe a system with perfect accessibility along with distinguished actions for that system. In the prescription of these mathematical objects, we will invoke notation and terminology introduced in Section 1, Chapter I and Section 1, Chapter II which provide useful points of contact with our study of homogeneous fluid bodies.

Definition 1.1. A *homogeneous fluid body with viscosity* $\mathscr{V}$ (or, more briefly, a *viscous body*) is prescribed by giving a *state space* $\Sigma_{\mathscr{V}}$, consisting of *states* $\sigma = (V, \theta, r)$, a collection $\Pi_{\mathscr{V}}$ of *process generators* π_t, a *pressure function* $\not{p}$, two *latent heat functions* $\tilde{\lambda}$ and $\tilde{\mu}$, and a *specific heat function* $\mathscr{s}$ which satisfy the following conditions:

($\mathscr{V}$1) $\Sigma_{\mathscr{V}}$ is of the form $\Sigma_{\mathscr{F}} \times \mathbb{R}$, with $\Sigma_{\mathscr{F}}$ an open, convex subset of $\mathbb{R}^{++} \times \mathbb{R}^{++}$;

($\mathscr{V}$2) $\not{p}$, $\tilde{\lambda}$, $\tilde{\mu}$, and σ are real-valued and continuously differentiable on $\Sigma_{\mathscr{V}}$;

($\mathscr{V}$3) throughout $\Sigma_{\mathscr{V}}$, $\partial \not{p}/\partial V$ is negative, σ is positive, and $\tilde{\lambda}$ does not vanish;

($\mathscr{V}$4) each process generator is a piecewise smooth function π_t from $[0, t)$ into $\mathbb{R}^2$ whose second component is a piecewise monotone function of τ and with t a positive real number.

We call V the *volume*, θ the *temperature* (with respect to a preassigned ideal gas scale $\varphi_{\mathscr{G}}$ which we take to be the same for all viscous bodies to be considered here), and r the *rate of expansion* of the body in the state (V, θ, r). Condition ($\mathscr{V}$1) permits us to visualize $\Sigma_{\mathscr{V}}$ as an unbounded cylinder, each of whose cross sections $r = \text{const.}$ is a translate of the set $\Sigma_{\mathscr{F}}$ lying in the first quadrant of the V-θ plane as shown in Figure 52. The reader will notice the resemblance between items ($\mathscr{V}$1)–($\mathscr{V}$3) above and items ($\mathscr{F}$1)–($\mathscr{F}$3) in the Definition 1.1, Chapter I, of a homogeneous fluid body. In fact, the set $\Sigma_{\mathscr{F}}$ in ($\mathscr{V}$1) may be regarded as the state space of a homogeneous fluid body $\mathscr{F}$ and, for each choice of r, the functions

$$(V,\theta) \mapsto \not{p}(V,\theta,r),$$
$$(V,\theta) \mapsto \tilde{\lambda}(V,\theta,r),$$
$$(V,\theta) \mapsto \sigma(V,\theta,r)$$

satisfy conditions ($\mathscr{F}$2) and ($\mathscr{F}$3) for a homogeneous fluid body. For the moment, we may think of the function $\tilde{\mu}$ not appearing in the definition of a homogeneous fluid body as an object corresponding to the variable r, also not appearing in that definition.

Condition ($\mathscr{V}$4) has no counterpart in the definition of a homogeneous fluid body, and it permits us to describe the pair $(\Sigma_{\mathscr{V}}, \Pi_{\mathscr{V}})$ as a system with perfect accessibility. This can be done in a manner analogous to the way in

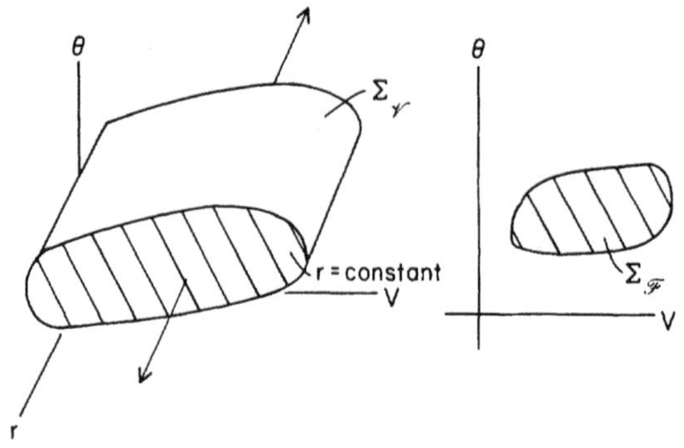

Figure 52.

which homogeneous fluid bodies (in Section 1 of Chapter II) and viscous filaments (in Section 3 of Chapter VI) were so described. For example, if π_t is a process generator, then we define $\mathscr{D}(\pi_t)$ to be the set of states $\sigma=(V,\theta,r)$ such that the differential equation

$$(V^{\cdot}(\tau),\theta^{\cdot}(\tau))=\pi_t(\tau)$$

with initial condition

$$(V(0),\theta(0))=(V,\theta)$$

has a solution $\tau\mapsto(V(\tau),\theta(\tau))$ lying entirely in $\Sigma_{\mathscr{F}}$ for $\tau\in[0,t]$; $\rho_{\pi_t}\sigma$ is then defined to be the state $(V(t),\theta(t),V^{\cdot}(t-))$. Each process (π_t,σ) determines not only an oriented curve $\Gamma(\pi_t,\sigma)$ in $\Sigma_{\mathscr{V}}$ parameterized by $\tau\mapsto(V(\tau),\theta(\tau),V^{\cdot}(\tau-))$, but also an oriented curve $\Gamma_{\text{eq}}(\pi_t,\sigma)$ in $\Sigma_{\mathscr{F}}\times\{0\}$ parameterized by $\tau\mapsto(V(\tau),\theta(\tau),0)$; the curve $\Gamma_{\text{eq}}(\pi_t,\sigma)$ is the projection of $\Gamma(\pi_t,\sigma)$ onto the plane $r=0$ and is called the *equilibrium path determined by the process* (π_t,σ). Using arguments similar to ones presented in Section 1 of Chapter II and Section 3 of Chapter VI, one can show that $(\Sigma_{\mathscr{V}},\Pi_{\mathscr{V}})$ is a system with perfect accessibility.

We are now in a position to describe viscous bodies as thermodynamical systems by defining certain actions for $(\Sigma_{\mathscr{V}},\Pi_{\mathscr{V}})$ in terms of the functions $\not{p}$, $\tilde{\lambda}$, $\tilde{\mu}$, and $\not{\sigma}$. The first action assigns to each process (π_t,σ) a number $\mathrm{W}_{\mathscr{V}}(\pi_t,\sigma)$ given by

$$\mathrm{W}_{\mathscr{V}}(\pi_t,\sigma):=\int_{\Gamma(\pi_t,\sigma)}\not{p}(V,\theta,r)\,dV$$

$$:=\int_0^t\not{p}(V(\tau),\theta(\tau),V^{\cdot}(\tau-))V^{\cdot}(\tau-)\,d\tau \qquad (1.1)$$

and called the *work done by the viscous body during the process* (π_t,σ). Next we prescribe an *accumulation function* $(\pi_t,\sigma,\mathrm{L})\mapsto H_{\mathscr{V}}(\pi_t,\sigma,\mathrm{L})$ which assigns to each process (π_t,σ) and hotness level L the number

$$H_{\mathscr{V}}(\pi_t,\sigma,\mathrm{L}):=\int_{\Gamma(\pi_t,\sigma,\mathrm{L})}\tilde{\lambda}(V,\theta,r)\,dV+\not{\sigma}(V,\theta,r)\,d\theta+\tilde{\mu}(V,\theta,r)\,dr$$

$$:=\int_{\ell(\pi_t,\sigma,\mathrm{L})}\big[\tilde{\lambda}(V(\tau),\theta(\tau),V^{\cdot}(\tau-))V^{\cdot}(\tau-)$$

$$+\not{\sigma}(V(\tau),\theta(\tau),V^{\cdot}(\tau-))\theta^{\cdot}(\tau)$$

$$+\tilde{\mu}(V(\tau),\theta(\tau),V^{\cdot}(\tau-))V^{\cdot\cdot}(\tau)\big]\,d\tau, \qquad (1.2)$$

with

$$\ell(\pi_t,\sigma,\mathrm{L}):=\{\tau\in[0,t]\,|\,\theta(\tau)\leqslant\varphi_{\mathscr{G}}(\mathrm{L})\} \qquad (1.3)$$

and with $\Gamma(\pi_t,\sigma,\mathrm{L})$ the portion of the curve $\Gamma(\pi_t,\sigma)$ on which θ does not exceed $\varphi_{\mathscr{G}}(\mathrm{L})$. (We write $\dot{V}(\tau)$ for $\dot{V}(\tau-)$ in the formulae (1.1) and (1.2) from now on.) We call $H_{\mathscr{V}}(\pi_t, \sigma, \mathrm{L})$ the *net heat gained by the viscous body* $\mathscr{V}$ *in the process* (π_t, σ) *at or below the hotness level* L. Relations (1.1) and (1.2) are the counterparts of (1.1) and (1.2), Chapter I, in our study of homogeneous fluid bodies. Of course, in the present relation (1.2), we specify not only the net heat gained by the viscous body in the process (π_t, σ), but also the net heat gained at or below each hotness level L. This is in accord with our desire to describe a viscous body as a thermodynamical system both in the sense of Chapter III and of Chapter IV, and the latter requires that an accumulation function be prescribed for the system.

It is important to note that $\mathrm{W}_{\mathscr{V}}$ and $H_{\mathscr{V}}$ are *not* rate-independent, as are the work and heat gained for a homogeneous fluid body $\mathscr{F}$. This fact can be understood simply by noting that the integrals with respect to τ in (1.1) and (1.2) contain terms which are non-linear in $\dot{V}$. Moreover, the term containing the second derivative $\ddot{V}$ in (1.2) has the effect of causing the path $\Gamma(\pi_t, \sigma, \mathrm{L})$ actually to change when $\tau \mapsto V(\tau)$ is replaced by a rescaling $\tau \mapsto V(\gamma(\tau))$ with γ increasing. The *accumulation integral* $\mathrm{I}_{\mathscr{V}}$, given by

$$\mathrm{I}_{\mathscr{V}}(\pi_t, \sigma) := \int_0^\infty H_{\mathscr{V}}(\pi_t, \sigma, \varphi_{\mathscr{G}}^{-1}(\theta))\theta^{-2}\, d\theta, \tag{1.4}$$

also is rate-dependent; this fact follows from the formula

$$\int_0^\infty H_{\mathscr{V}}(\pi_t, \sigma, \varphi_{\mathscr{G}}^{-1}(\theta))\theta^{-2}\, d\theta$$

$$= \int_{\Gamma(\pi_t, \sigma)} \frac{\tilde{\lambda}(V, \theta, r)}{\theta} dV + \frac{\tilde{\kappa}(V, \theta, r)}{\theta} d\theta + \frac{\tilde{\mu}(V, \theta, r)}{\theta} dr \tag{1.5}$$

(which, in turn, follows from the arguments used to verify (3.5), Chapter IV). Indeed, (1.4) and (1.5) tell us that $\mathrm{I}_{\mathscr{V}}(\pi_t, \sigma)$ can be expressed in a form similar to $\mathrm{W}_{\mathscr{V}}(\pi_t, \sigma)$ and $H_{\mathscr{V}}(\pi_t, \sigma, \mathrm{L})$ so that the arguments given to establish the rate-dependence of $\mathrm{W}_{\mathscr{V}}$ and $H_{\mathscr{V}}$ apply as well to $\mathrm{I}_{\mathscr{V}}$. The rate-dependence of these quantities will be studied in more detail in the following sections, where their behavior under retardations of processes will play an important role in our analysis of consequences of the laws of thermodynamics.

Methods which were employed several times in Chapters II and VI permit us to call the functions $(\pi_t, \sigma) \mapsto \mathrm{W}_{\mathscr{V}}(\pi_t, \sigma)$, $(\pi_t, \sigma) \mapsto H_{\mathscr{V}}(\pi_t, \sigma, \mathrm{L})$, and $(\pi_t, \sigma) \mapsto \mathrm{I}_{\mathscr{V}}(\pi_t, \sigma)$ actions for the system with perfect accessibility $(\Sigma_{\mathscr{V}}, \Pi_{\mathscr{V}})$. Consequently, we have shown here that a homogeneous body with viscosity is a thermodynamical system in both the first and second senses in which this concept has been used, and we can now study these bodies from the point of view of our modern approach to thermodynamics.

2. Consequences of the First Law

The First Law for a viscous body $\mathscr{V}$ is the assertion that, *if the net heat gained in a cycle of* $\mathscr{V}$ *vanishes, then so must the work done by* $\mathscr{V}$ *in that cycle.* Our study of the First Law in earlier chapters can be summarized by means of Corollary 3.1, Chapter V: *Let* $\mathscr{G}$ *be an ideal gas such that* $\mathscr{V}$ *and* $\mathscr{G}$ *preserve the First Law with respect to the product operation, and let* $(V^\circ, \theta^\circ, r^\circ)$ *be a state of* $\mathscr{V}$. *It follows that* $\mathscr{V}$ *obeys the First Law if and only if there is exactly one energy function* E° *for* $\mathscr{V}$ *which vanishes at* $(V^\circ, \theta^\circ, r^\circ)$, i.e., *there is exactly one function* $\mathrm{E}^\circ\colon \Sigma_{\mathscr{V}} \to \mathbb{R}$ *such that*

$$\mathrm{E}^\circ(V^\circ, \theta^\circ, r^\circ) = 0 \tag{2.1}$$

and

$$\mathrm{E}^\circ(V_2, \theta_2, r_2) - \mathrm{E}^\circ(V_1, \theta_1, r_1) = \left(\frac{\mathrm{R}}{\lambda}\right) H_{\mathscr{V}}(\pi_t, (V_1, \theta_1, r_1)) - \mathrm{W}_{\mathscr{V}}(\pi_t, (V_1, \theta_1, r_1)) \tag{2.2}$$

for every pair of states (V_1, θ_1, r_1), (V_2, θ_2, r_2) *of* $\mathscr{V}$ *and process generator* π_t *satisfying*

$$\rho_{\pi_t}(V_1, \theta_1, r_1) = (V_2, \theta_2, r_2). \tag{2.3}$$

It is convenient in what follows to write J for R/λ, the ratio of the constants associated with the ideal gas $\mathscr{G}$, and to assume that each viscous body under consideration together with $\mathscr{G}$ preserves the First Law under the product operation.

Guided by the results in Sections 3 and 5 of Chapter VI, we study the behavior of the relation (2.2) under retardations of processes. We already remarked at the beginning of Chapter I that it is customary to regard evolution of a body through a collection of spatially homogeneous states as an idealized description of evolution which is slow in an appropriate sense. Our first task here will be to show that the energy function for a homogeneous body $\mathscr{V}$ with viscosity is determined by a homogeneous fluid body $\mathscr{F}$ which approximates $\mathscr{V}$ in slow processes.

Theorem 2.1. *Let* t, V_1, *and* V_2 *be positive numbers and let* $\tau \mapsto (V(\tau), \theta(\tau))$ *be a continuously differentiable function from* $[0, t]$ *into* $\Sigma_{\mathscr{F}}$ (*the open, convex subset of* $\mathbb{R}^{++} \times \mathbb{R}^{++}$ *associated with the state space* $\Sigma_{\mathscr{V}} = \Sigma_{\mathscr{F}} \times \mathbb{R}$ *of a given viscous body*), *with* $\tau \mapsto V(\tau)$ *satisfying*

$$V(0) = V_1 \quad \text{and} \quad V(t) = V_2. \tag{2.4}$$

If E° *is an energy function for* $\mathscr{V}$, *then for each* r_1, r_2 *in* $\mathbb{R}$, *there holds*

$$\begin{aligned}\mathrm{E}^\circ(V_2, \theta_2, r_2) &- \mathrm{E}^\circ(V_1, \theta_1, r_1) \\ &= \mathrm{J}\int_0^t [\tilde{\lambda}(V(\tau), \theta(\tau), 0) V^{\boldsymbol{\cdot}}(\tau) + \mathscr{c}(V(\tau), \theta(\tau), 0)\theta^{\boldsymbol{\cdot}}(\tau)]\, d\tau \\ &\quad - \int_0^t \mathscr{p}(V(\tau), \theta(\tau), 0) V^{\boldsymbol{\cdot}}(\tau)\, d\tau,\end{aligned} \tag{2.5}$$

where θ_1 and θ_2 denote $\theta(0)$ and $\theta(t)$, respectively. It follows that E° *is independent of the third variable r, the rate of expansion, and there holds*

$$\frac{\partial \mathrm{E}^\circ}{\partial V}(V,\theta,r) = \mathrm{J}\tilde{\lambda}(V,\theta,0) - p(V,\theta,0), \tag{2.6}$$

$$\frac{\partial \mathrm{E}^\circ}{\partial \theta}(V,\theta,r) = \mathrm{J}c(V,\theta,0) \tag{2.7}$$

for every state (V,θ,r) of $\mathscr{V}$.

PROOF. Relation (2.5) is similar to (3.13), Chapter VI, and the proof of (3.13) is easily modified to yield a proof of (2.5). In fact, we have only to apply the same retardation described in (3.14), Chapter VI, for a function $\tau \mapsto \ell(\tau)$ to the given function $\tau \mapsto V(\tau)$ without alteration of the function $\tau \mapsto \theta(\tau)$. By this means, it is immediate from the proof of Lemma 3.1, Chapter VI, that the expression

$$\int_0^t \{[\mathrm{J}\tilde{\lambda}(V(\tau),\theta(\tau),V^\cdot(\tau)) - p(V(\tau),\theta(\tau),V^\cdot(\tau))]V^\cdot(\tau)$$
$$+ \mathrm{J}c(V(\tau),\theta(\tau),V^\cdot(\tau))\theta^\cdot(\tau)\}\,d\tau$$

on the right-hand side of (2.2) tends to the right-hand side of (2.5) in the limit under retardations. The remaining term

$$\int_0^t \mathrm{J}\tilde{\mu}(V(\tau),\theta(\tau),V^\cdot(\tau))V^{\cdot\cdot}(\tau)\,d\tau$$

on the right-hand side of (2.2) tends to zero as n tends to infinity because the factor $V^{\cdot\cdot}(\tau)$ tends to zero more rapidly under retardation than does $V^\cdot(\tau)$. Of course, the left-hand side of (2.2) is unchanged under the retardations employed in this proof, and this establishes (2.5). The relations (2.6) and (2.7) as well as the lack of dependence of E° upon r follow immediately from (2.5) and familiar theorems from Vector Analysis. □

It is clear from Theorem 2.1 that as far as the determination of E° is concerned, one may restrict the functions p, $\tilde{\lambda}$, and c to the plane $r = 0$, i.e., $\Sigma_{\mathscr{V}}$ can be replaced by $\Sigma_{\mathscr{F}}$, and one may take $\tilde{\mu}$ to be zero. This amounts to replacing $\mathscr{V}$ by a homogeneous fluid body $\mathscr{F}$ with state space $\Sigma_{\mathscr{F}}$ and response functions $(V,\theta) \mapsto p(V,\theta,0)$, $(V,\theta) \mapsto \tilde{\lambda}(V,\theta,0)$, and $(V,\theta) \mapsto c(V,\theta,0)$. Because this replacement is justified by taking processes of $\mathscr{V}$ in which the rate of expansion tends to zero, we may regard $\mathscr{F}$ as an approximation to $\mathscr{V}$ for slow processes of the viscous body.

It is now possible to give conditions on the response functions p, $\tilde{\lambda}$, $\tilde{\mu}$, and c of a viscous body $\mathscr{V}$ which are both necessary and sufficient in order that $\mathscr{V}$ obey the First Law.

Theorem 2.2. *A viscous body $\mathscr{V}$ obeys the First Law if and only if the functions $\not p$, $\tilde{\lambda}$, $\tilde{\mu}$, and σ obey the following relations for every (V,θ,r) in $\Sigma_{\mathscr{V}}$:*

$$J\tilde{\lambda}(V,\theta,r)-\not p(V,\theta,r)=J\tilde{\lambda}(V,\theta,0)-\not p(V,\theta,0), \tag{2.8}$$

$$\sigma(V,\theta,r)=\sigma(V,\theta,0), \tag{2.9}$$

$$\tilde{\mu}(V,\theta,r)=0, \tag{2.10}$$

$$\frac{\partial}{\partial\theta}\left[J\tilde{\lambda}(V,\theta,0)-\not p(V,\theta,0)\right]=\frac{\partial}{\partial V}\left[J\sigma(V,\theta,0)\right]. \tag{2.11}$$

PROOF. If $\mathscr{V}$ obeys the First Law, then there is an energy function E° satisfying (2.1) and (2.2), and Theorem 2.1 tells us that, not only does E° satisfy (2.2), but also (2.5) is satisfied, so that in *every* process of $\mathscr{V}$ there holds

$$\begin{aligned}&\int_0^t\{[J\tilde{\lambda}(V(\tau),\theta(\tau),V^{\cdot}(\tau))-\not p(V(\tau),\theta(\tau),V^{\cdot}(\tau))]V^{\cdot}(\tau)\\&\quad+\sigma(V(\tau),\theta(\tau),V^{\cdot}(\tau))\theta^{\cdot}(\tau)+\tilde{\mu}(V(\tau),\theta(\tau),V^{\cdot}(\tau))V^{\cdot\cdot}(\tau)\}\,d\tau\\&=\int_0^t\{[J\tilde{\lambda}(V(\tau),\theta(\tau),0)-\not p(V(\tau),\theta(\tau),0)]V^{\cdot}(\tau)\\&\quad+\sigma(V(\tau),\theta(\tau),0)\theta^{\cdot}(\tau)\}\,d\tau.\end{aligned} \tag{2.12}$$

By taking a process with initial state (V,θ,r) chosen arbitrarily, and with $V^{\cdot}(\tau)$ identically equal to r and $\theta^{\cdot}(\tau)$ identically zero, (2.12) yields

$$\begin{aligned}&\int_0^t\{J\tilde{\lambda}(V(\tau),\theta,r)-\not p(V(\tau),\theta,r)\}r\,d\tau\\&\quad=\int_0^t\{J\tilde{\lambda}(V(\tau),\theta,0)-\not p(V(\tau),\theta,0)\}r\,d\tau.\end{aligned}$$

Dividing this relation by t and letting t decrease to zero, we find that

$$\{J\tilde{\lambda}(V,\theta,r)-\not p(V,\theta,r)\}r=\{J\tilde{\lambda}(V,\theta,0)-\not p(V,\theta,0)\}r.$$

If r is not zero, then the state (V,θ,r) obeys (2.8). Of course, if r is zero, then (2.8) clearly is satisfied. Because we have verified (2.8) for all states, (2.12) reduces to

$$\begin{aligned}&\int_0^t\{\sigma(V(\tau),\theta(\tau),V^{\cdot}(\tau))\theta^{\cdot}(\tau)+\tilde{\mu}(V(\tau),\theta(\tau),V^{\cdot}(\tau))V^{\cdot\cdot}(\tau)\}\,d\tau\\&\quad=\int_0^t\sigma(V(\tau),\theta(\tau),0)\theta^{\cdot}(\tau)\,d\tau,\end{aligned} \tag{2.13}$$

which again must hold for all processes. For a process with arbitrarily chosen initial state (V,θ,r) and with $V^{\cdot}(\tau)$ identically equal to r and $\theta^{\cdot}(\tau)$

identically equal to a number γ, (2.13) yields

$$\int_0^t \mathscr{d}(V(\tau),\theta(\tau),r)\gamma\, d\tau = \int_0^t \mathscr{d}(V(\tau),\theta(\tau),0)\gamma\, d\tau,$$

and this implies that, for all choices of γ,

$$\mathscr{d}(V,\theta,r)\gamma = \mathscr{d}(V,\theta,0)\gamma.$$

The relation (2.9) follows immediately, and (2.13) becomes

$$\int_0^t \tilde{\mu}(V(\tau),\theta(\tau),V^{\cdot}(\tau))V^{\cdot\cdot}(\tau)\, d\tau = 0. \tag{2.14}$$

For a process starting at (V,θ,r), division of (2.14) by t and taking the limit as t decreases to zero gives

$$\tilde{\mu}(V,\theta,V^{\cdot}(0+))V^{\cdot\cdot}(0+) = 0.$$

However, $V^{\cdot}(0+)$ and $V^{\cdot\cdot}(0+)$, the limits of $V^{\cdot}(\tau)$ and $V^{\cdot\cdot}(\tau)$ as τ decreases to zero, can be chosen arbitrarily, and this fact implies the relation (2.10). Moreover, (2.11) is an immediate consequence of (2.6), (2.7), and the fact that E° is twice differentiable. Conversely, if (2.8) through (2.11) hold, then the relation

$$\begin{aligned} JH_{\mathscr{V}}(\pi_t,(V,\theta,r)) - W_{\mathscr{V}}(\pi_t,(V,\theta,r)) &= \int_0^t \{[J\tilde{\lambda}(V(\tau),\theta(\tau),0) - \not{p}(V(\tau),\theta(\tau),0)]V^{\cdot}(\tau) \\ &\quad + J\mathscr{d}(V(\tau),\theta(\tau),0)\theta^{\cdot}(\tau)\}\, d\tau \end{aligned}$$

holds for every process $(\pi_t,(V,\theta,r))$ of $\mathscr{V}$, and this relation can be written in the form

$$\begin{aligned} JH_{\mathscr{V}}(\pi_t,(V,\theta,r)) - W_{\mathscr{V}}(\pi_t,(V,\theta,r)) &= \int_{\Gamma_{eq}(\pi_t,(V,\theta,r))} (J\tilde{\lambda}(V,\theta,0) - \not{p}(V,\theta,0))\, dV + J\mathscr{d}(V,\theta,0)\, d\theta. \end{aligned} \tag{2.15}$$

However, (2.11) tells us that $(J\tilde{\lambda}(V,\theta,0) - \not{p}(V,\theta,0))\, dV + J\mathscr{d}(V,\theta,0)\, d\theta$ is an exact differential on $\Sigma_{\mathscr{F}}$, so that the right-hand side of (2.15) vanishes whenever $\Gamma_{eq}(\pi_t,(V,\theta,r))$ is a closed curve. If $(\pi_t,(V,\theta,r))$ is a cycle of $\mathscr{V}$, then $\Gamma_{eq}(\pi_t,(V,\theta,r))$ *is* a closed curve, and we conclude from (2.15) that $JH_{\mathscr{V}} - W_{\mathscr{V}}$ vanishes on the cycles of $\mathscr{V}$, i.e., $\mathscr{V}$ satisfies Joule's relation. As we pointed out in Chapter III, Joule's relation implies that the First Law holds, and this completes the proof. □

Results of the type given in Theorem 2.2 are the ultimate goal of a thermodynamical analysis of any class of physical systems, because they express the entire content of a basic law in terms of functions which are central to the description of the systems. Relations (2.8), (2.9), and (2.11) combine to yield:

$$\frac{\partial}{\partial\theta}(\mathrm{J}\tilde{\lambda}-\mathscr{p})=\mathrm{J}\frac{\partial\mathscr{c}}{\partial V}. \tag{2.16}$$

[For a homogeneous fluid body, Theorem 2.1 of Chapter I tells us that (2.16) alone expresses the entire content of the First Law.] For viscous bodies, the extent to which $H_{\mathscr{Y}}(\pi_t,(V,\theta,r))$ can be influenced by the values of the rate of expansion during the process $(\pi_t,(V,\theta,r))$ is made precise by means of (2.8) through (2.10), for (2.9) and (2.10) imply that

$$\begin{aligned}H_{\mathscr{Y}}(\pi_t,(V,\theta,r))=\int_0^t\{&\tilde{\lambda}(V(\tau),\theta(\tau),V^{\cdot}(\tau))V^{\cdot}(\tau)\\&+\mathscr{c}(V(\tau),\theta(\tau),0)\theta^{\cdot}(\tau)\}\,d\tau,\end{aligned}$$

and (2.8) then yields:

$$\begin{aligned}&H_{\mathscr{Y}}(\pi_t,(V,\theta,r))\\&\quad=\int_{\Gamma_{\mathrm{eq}}(\pi_t,(V,\theta,r))}\tilde{\lambda}(V,\theta,0)\,dV+\mathscr{c}(V,\theta,0)\,d\theta\\&\qquad+\frac{1}{\mathrm{J}}\int_0^t\{\mathscr{p}(V(\tau),\theta(\tau),V^{\cdot}(\tau))-\mathscr{p}(V(\tau),\theta(\tau),0)\}V^{\cdot}(\tau)\,d\tau.\end{aligned} \tag{2.17}$$

As was suggested in our discussion of viscous filaments in Section 3 of Chapter VI, we call the integral

$$\int_0^t\{\mathscr{p}(V(\tau),\theta(\tau),V^{\cdot}(\tau))-\mathscr{p}(V(\tau),\theta(\tau),0)\}V^{\cdot}(\tau)\,d\tau \tag{2.18}$$

the work done by the viscous body through extra forces, and (2.17) is then the assertion that *the net heat gained by a viscous body in any process equals the net heat gained by the approximating homogeneous fluid body plus an amount of heat equivalent to the work done by the viscous body through the extra forces*. A result analogous to (2.17) was encountered earlier in our study of phase transitions of homogeneous filaments; in fact, relation (5.10) of Chapter VI is of the same form as (2.17). [In (5.10), ϕ_{v} is the force exerted *on* the filament, so that the integral in (5.10) is the work done *on* the filament through extra forces; this explains why a minus sign appears in (5.10) in place of the plus sign in (2.17).] From our work in Chapter VI on consequences of the Second Law for viscous filaments, we might expect that the Second Law should give us information about the work done by a viscous body through the extra forces. The next section is devoted to a study of questions of this type.

3. Consequences of the Second Law

The Second Law for a viscous body $\mathscr{V}$ is the assertion that, *if the net heat gained at or below every hotness level during a cycle is non-negative, then the net heat gained in the cycle is zero*. As a result of the analysis presented in Chapters IV and V, we have from Corollary 2.1, Chapter V: *Let $\mathscr{G}$ be an ideal gas such that $\mathscr{V}$ and $\mathscr{G}$ preserve the Second Law with respect to the product operation. It follows that $\mathscr{V}$ obeys the Second Law if and only if there is an entropy function for $\mathscr{V}$, i.e., there exists a real-valued function* S *on $\Sigma_{\mathscr{V}}$ satisfying*

$$\int_0^{\infty} H_{\mathscr{V}}(\pi_t, V_1, \theta_1, r_1, \varphi_{\mathscr{G}}^{-1}(\theta))\theta^{-2}\,d\theta \leqslant \mathrm{S}(V_2, \theta_2, r_2) - \mathrm{S}(V_1, \theta_1, r_1) \tag{3.1}$$

whenever (V_2, θ_2, r_2) equals $\rho_{\pi_t}(V_1, \theta_1, r_1)$.

We assume once and for all in this section that each viscous body under consideration, together with a preassigned ideal gas $\mathscr{G}$, preserves the Second Law with respect to the product operation. We then may say that the Second Law is equivalent to the existence of an entropy function. Although we were able to prove in Chapter V that every thermodynamical system has *at most* one normalized energy function, there is no corresponding uniqueness result concerning entropy functions. (See Theorem 2.3 of Chapter V for conditions equivalent to uniqueness of normalized entropy functions.) Nevertheless, the special features of viscous bodies will turn out to guarantee uniqueness of normalized entropy functions, just as was the case for normalized Helmholtz free energy functions for the viscous filaments studied in Section 3 of Chapter VI. The next result corresponds to Theorem 2.1 of this chapter, in that it shows that each entropy function for a viscous body $\mathscr{V}$ is determined by a homogeneous fluid body $\mathscr{F}$ which approximates $\mathscr{V}$ for sufficiently slow processes.

Theorem 3.1. *Let $\mathscr{V}$ be a viscous body with state space $\Sigma_{\mathscr{V}} = \Sigma_{\mathscr{F}} \times \mathbb{R}$, let t, V_1, and V_2 be positive numbers, and let $\tau \mapsto (V(\tau), \theta(\tau))$ be a continuously differentiable function from $[0, t]$ into $\Sigma_{\mathscr{F}}$ such that*

$$V(0) = V_1 \quad \text{and} \quad V(t) = V_2. \tag{3.2}$$

If S *is an entropy function for $\mathscr{V}$, then for each r_1, r_2 in $\mathbb{R}$ there holds*

$$\mathrm{S}(V_2, \theta_2, r_2) - \mathrm{S}(V_1, \theta_1, r_1) = \int_0^t \left[\frac{\tilde{\lambda}(V(\tau), \theta(\tau), 0)}{\theta(\tau)} V^{\cdot}(\tau) + \frac{\tilde{\sigma}(V(\tau), \theta(\tau), 0)}{\theta(\tau)} \theta^{\cdot}(\tau) \right] d\tau. \tag{3.3}$$

(Here, θ_1 and θ_2 denote $\theta(0)$ and $\theta(t)$, respectively.) Consequently, S *is*

independent of the variable r and satisfies

$$\frac{\partial S}{\partial V}(V,\theta,r)=\frac{\tilde{\lambda}(V,\theta,0)}{\theta}, \tag{3.4}$$

$$\frac{\partial S}{\partial \theta}(V,\theta,r)=\frac{\sigma(V,\theta,0)}{\theta} \tag{3.5}$$

for every state (V,θ,r) *of* $\mathscr{V}$.

PROOF. The proof of (3.3) is similar to that of (2.5) and of (3.13), Chapter VI. The only step required here and not in the earlier proofs is that of showing that the integral in the left-hand member of (3.1), i.e., the accumulation integral for $\mathscr{V}$, can be written in a form similar to the right-hand sides of (1.1) and (1.2). However, the relation (1.5) supplies this needed step. □

Theorem 3.1 tells us that each entropy function for $\mathscr{V}$ is determined by a homogeneous fluid body $\mathscr{F}$ with state space $\Sigma_{\mathscr{F}}$, latent heat function $(V,\theta)\mapsto\tilde{\lambda}(V,\theta,0)$, and specific heat function $(V,\theta)\mapsto\sigma(V,\theta,0)$. In fact, this theorem shows that the partial derivatives of any two entropy functions agree throughout the connected set $\Sigma_{\mathscr{V}}$, and we therefore have the following result.

Corollary 3.1. *Any two entropy functions for a viscous body differ by at most a constant. In particular, for each state* $(V^\circ,\theta^\circ,r^\circ)$ *of the body, there is at most one entropy function* S° *which vanishes at* $(V^\circ,\theta^\circ,r^\circ)$.

We continue our study of the Second Law with a result similar to Theorem 2.2.

Theorem 3.2. *A viscous body* $\mathscr{V}$ *obeys the Second Law if and only if the functions* $\tilde{\lambda}$, $\tilde{\mu}$, *and* σ *obey the relation* (3.6)–(3.9) *throughout* $\Sigma_{\mathscr{V}}$:

$$(\tilde{\lambda}(V,\theta,r)-\tilde{\lambda}(V,\theta,0))r\leqslant 0, \tag{3.6}$$

$$\sigma(V,\theta,r)=\sigma(V,\theta,0), \tag{3.7}$$

$$\tilde{\mu}(V,\theta,r)=0, \tag{3.8}$$

$$\frac{\partial}{\partial\theta}\left[\frac{\tilde{\lambda}(V,\theta,0)}{\theta}\right]=\frac{\partial}{\partial V}\left[\frac{\sigma(V,\theta,0)}{\theta}\right]. \tag{3.9}$$

PROOF. The proof of this theorem differs from that of Theorem 2.2 only because the relation (2.2) is an equation and (3.1) is an inequality. However, a step by step examination of the proof of Theorem 2.2 easily reveals the truth of the assertions made in the present theorem. □

We now suppose that a viscous body $\mathscr{V}$ obeys *both* the First and Second Laws, so that both sets of relations (2.8)–(2.11) and (3.6)–(3.9) hold. It follows from (2.11), (3.9), and Theorem 3.2 of Chapter I that the approximating homogeneous fluid body $\mathscr{F}$ obeys both the First and Second Laws. Moreover, (2.8) and (3.6) yield the inequality

$$[\not p(V,\theta,r)-\not p(V,\theta,0)]\,r\leqslant 0, \tag{3.10}$$

which in turn tells us that the integral in (2.18) is not positive, i.e., *the First and Second Laws imply that the work done by the viscous body through extra forces is not positive*. Using this result and (2.17), we conclude that

$$H_{\mathscr{V}}(\pi_t,(V,\theta,r))\leqslant\int_{\Gamma_{\mathrm{eq}}(\pi_t,(V,\theta,r))}\tilde{\lambda}(V,\theta,0)\,dV+\not c(V,\theta,0)\,d\theta \tag{3.11}$$

for every process $(\pi_t,(V,\theta,r))$ of $\mathscr{V}$, i.e., *in any process the net heat gained by the viscous body $\mathscr{V}$ cannot exceed the net heat gained by the approximating homogeneous fluid body $\mathscr{F}$*. Relation (3.10) may also be translated into a similar statement: *in any process, the work done by the viscous body cannot exceed the work done by the approximating homogeneous fluid body.*

If we return to a principal topic in classical thermodynamics, the study of the efficiency of heat engines, we obtain from (2.17) and the last italicized statement the following interesting result: *For any process $(\pi_t,(V,\theta,r))$ of $\mathscr{V}$ in which the heat absorbed is not zero, the efficiency of the process for the viscous body is no greater than the efficiency of the associated process $(\pi_t,(V,\theta))$ for the approximating homogeneous fluid body. The efficiency for the viscous body approaches that for the homogeneous fluid body when processes of the viscous body are retarded indefinitely.* Thus, the presence of viscous forces in a homogeneous body cannot improve the efficiency of a heat engine which uses that body as a working substance. For this reason, viscous forces are said to be dissipative (or wasteful) when it is desired to extract work from a substance by heating and cooling that substance. If one were only concerned with the actual amount of work extracted per unit amount of heat absorbed, so that the duration of processes could be as large as desired, then the dissipation could be made as small as desired by operating a heat engine sufficiently slowly. However, one generally needs to extract work from a substance within a preassigned interval of time, and the presence of viscous forces must be reckoned with in order to obtain realistic estimates of the efficiency of heat engines.

Comments and Suggestions for Further Reading

Much of the impetus for research on the mathematical foundations of thermodynamics during the past twenty years was created by the rapid development during the previous twenty years of the non-linear field theories of mechanics. That development provided a conceptual framework for the formulation of "constitutive equations," relations which distinguish one material from another. The study of a variety of new constitutive equations together with the view that the laws of thermodynamics should place restrictions on the new equations led to the discovery of systematic procedures for finding such restrictions. In a relatively short time, these procedures were widely used, although not always without reservation. In fact, the new procedures required that one use concepts such as absolute temperature, internal energy, and entropy in contexts where traditional thermodynamics could not in any obvious way provide justification. The need to interpret these concepts for materials lying outside the scope of classical treatments, particularly for materials whose present state can depend in a non-trivial way on the entire past history of the material, was the main force behind the modern work on the foundations of thermodynamics. Many of the developments in the non-linear field theories of mechanics during the twenty years following World War II are described in great depth in the article by Truesdell and Noll [1]. Included in that article are discussions of papers by Coleman and Noll [2], and Coleman and Mizel [3], which introduce a procedure for finding thermodynamical restrictions on constitutive equations, and a description of Coleman's thermodynamics of materials with memory [4], [5].

One phase of recent work on foundations has focused on justifying the use of internal energy and entropy for non-classical materials. This task requires the formulation of versions of the First and Second Laws that do not mention energy and entropy and that are sufficiently general to cover the non-classical materials. Moreover, these versions of the laws of thermodynamics should yield as theorems the existence of functions having the basic properties of energy and of entropy required for both the classical and the new applications. A decisive step in this area of research was made by Day [6], [7] in his treatment of the Second Law for materials with fading memory. Day's work—together with an important paper of Noll [8] in which concepts from mathematical systems theory were used to describe certain mechanical systems—led Coleman and Owen [9] to formulate a mathematical foundation for thermodynamics containing versions of the First and Second Laws and theorems on the existence of energy and entropy functions of the type described above. This research underlies the material presented in Chapters II and V of this book. The discussion of uniqueness of entropy functions in Chapter V is based not only on results of Coleman and Owen [9], but also on more recent, as yet unpublished, work of Serrin.

A second area of modern research is concerned with reformulating classical, verbal statements of the First and Second Laws of thermodynamics so as to make precise their physical and mathematical content and to clarify both the notion of mechanical equivalence of heat and the notion of absolute temperature. James Serrin's research [10], [11], [12] on accumulation functions for thermodynamical systems and Miroslav Šilhavý's research on heating measures [13], [14] accomplished these goals for the Second Law and the notion of absolute temperature. Serrin's work brought out the importance of the concept of a "union" (or product) of thermodynamical systems and forms the basis of much of Chapter IV. The idea that the First Law can be formulated in a manner which yields Joule's relation as a consequence and that the First and Second Laws can be given parallel treatments is due to Šilhavý [15]. These ideas have contributed much to the material in Chapter III of this book. Further development of the modern approach to the Second Law can be found in the papers by Feinberg and Lavine [31] and Coleman, Owen, and Serrin [32].

A third area of modern research, one which is not represented in this text, concerns the formulation and analysis of the laws of thermodynamics as they apply to three dimensional, continuous bodies whose state can vary from point to point. In particular, one strives to find global forms of the laws that apply to material bodies and that are equivalent to local forms, usually partial differential equations, required to hold at almost all points of the body. A study of this area requires more advanced tools from analysis and geometry than do the first two areas and so goes beyond the scope of this book. References [16]–[21] contain important contributions to this research.

The forthcoming, second edition of Clifford Truesdell's *Rational Thermodynamics*, to be published by Springer-Verlag, will contain brief accounts of many interesting areas of modern research. Generally speaking, these accounts are less technical than most papers in the literature, and so provide good introductions to many subjects, some of which are not included in this book. References [22]–[27] provide background material for this text in the areas of mathematical analysis, calorimetry, and thermometry. (Some recent work on the concepts of hotness and temperature can be found in Reference [31] and in the paper of Fosdick and Rajagopal [33].) In references [28]–[30] the reader will find mentioned texts on thermodynamics which are written from a different point of view than the present one and which, in particular, go beyond applications to thermomechanics alone.

References

1. Truesdell, C., Noll, W. (1965). The Non-linear Field Theories of Mechanics. In: Encyclopedia of Physics, Vol. III/3. Berlin Heidelberg New York: Springer-Verlag.
2. Coleman, B. D., Noll, W. (1963). The thermodynamics of elastic materials with heat conduction and viscosity. Arch. Rational Mech. Anal. *13*, 167–178.
3. Coleman, B. D., Mizel, V. J. (1964). Existence of caloric equations of state in thermodynamics. J. Chem. Phys. *40*, 1116–1125.
4. Coleman, B. D. (1964). Thermodynamics of materials with memory. Arch. Rational Mech. Anal. *17*, 1–46.
5. Coleman, B. D. (1964). On thermodynamics, strain impulses, and viscoelasticity. Arch. Rational Mech. Anal. *17*, 230–254.
6. Day, W. A. (1969). A theory of thermodynamics for materials with memory. Arch. Rational Mech. Anal. *34*, 85–96.
7. Day, W. A. (1972). The Thermodynamics of Simple Materials with Fading Memory. Springer Tracts in Natural Philosophy, Vol. 22. Berlin Heidelberg New York: Springer-Verlag.
8. Noll, W. (1972). A new mathematical theory of simple materials. Arch. Rational Mech. Anal. *48*, 1–50.
9. Coleman, B. D., Owen, D. R. (1974). A mathematical foundation for thermodynamics. Arch. Rational Mech. Anal. *54*, 1–104.
10. Serrin, J. (1977). The concepts of thermodynamics. In: (1978) Contemporary Developments in Continuum Mechanics and Partial Differential Equations, G. M. de la Penha and L. A. Medeiros (eds.), pp. 411–451. Amsterdam: North Holland Co.
11. Serrin, J. (1979). Lectures on Thermodynamics. University of Naples.
12. Serrin, J. (1979). Conceptual analysis of the classical second laws of thermodynamics. Arch. Rational Mech. Anal. *70*, 355–371.
13. Šilhavý, M. (1978). On the Clausius Inequality. Abstracts, Czechoslovak Academy of Sciences and Škoda, National Corporation, Plzeň, p. 68.
14. Šilhavý, M. (1979). On the Clausius Inequality; published in revised form: (1983) On the Clausius Inequality. Arch. Rational Mech. Anal. *81*, 221–243.
15. Šilhavý, M. (1980). On measures, convex cones, and foundations of thermodynamics; I. Systems with vector-valued actions; II. Thermodynamic systems. Czech, J. Phys. *B 30*, 841–861, 961–991.
16. Noll, W. (1959). The foundations of classical mechanics in the light of recent advances in continuum mechanics. In: The Axiomatic Method with Special Reference to Geometry and Physics, 266–281. Amsterdam: North Holland Co.
17. Gurtin, M. E., Williams, W. O. (1967). An axiomatic foundation for continuum thermodynamics. Arch. Rational Mech. Anal. *26*, 83–117.

18. Gurtin, M. E., Mizel, V. J., Williams, W. O. (1968). On Cauchy's stress theorem. J. Math. Anal. Appl. *22*, 398–401.
19. Williams, W. O. (1969). On internal interactions and the concept of thermal isolation. Arch. Rational Mech. Anal. *34*, 245–258.
20. Gurtin, M. E., Williams, W. O. (1971). On the first law of thermodynamics. Arch. Rational Mech. Anal. *42*, 77–92.
21. Gurtin, M. E., Martins, L. C. (1976). Cauchy's theorem in classical physics. Arch. Rational Mech. Anal. *60*, 305–324.
22. Taylor, A. E. (1955). Advanced Calculus. Waltham Toronto London: Blaisdell.
23. Goldberg, R. R. (1964). Methods of Real Analysis. Waltham Toronto London: Blaisdell.
24. Fulks, W. (1969). Advanced Calculus (Second Edition). New York: John Wiley and Sons.
25. Hall, J. A. (1953). Fundamentals of Thermometry. London: Chapman and Hall, Ltd.
26. Brickwedde, F. G. (ed.) (1962). Temperature: Its Measurement and Control in Science and Industry, Vol. *3*, Part 1. London: Chapman and Hall, Ltd.
27. Middleton, W. E. Knowles (1966). A History of the Thermometer and Its Uses in Meteorology. Baltimore: Johns Hopkins Press.
28. Planck, Max (1927). Treatise on Thermodynamics (Fifth Edition). London: Longmans, Green and Co., Ltd.
29. Pippard, A. B. (1957). Elements of Classical Thermodynamics. Cambridge University Press.
30. Callen, H. B. (1960). Thermodynamics. New York London Sidney: John Wiley and Sons.
31. Feinberg, M., Lavine, R. (1983). Thermodynamics based on the Hahn-Banach Theorem: the Clausius Inequality. Arch. Rational Mech. Anal. *82*, 203–293; (Forthcoming) Thermodynamics based on the Hahn-Banach Theorem: the Clausius-Duhem Inequality.
32. Coleman, B. D., Owen, D. R., Serrin, J. (1981). The second law of thermodynamics for systems with approximate cycles. Arch. Rational Mech. Anal. *77*, 103–142.
33. Fosdick, R. L., Rajagopal, K. R. (1983). On the existence of a manifold for temperature. Arch. Rational Mech. Anal. *81*, 317–332.

Problems

Chapter I

1. Prove Theorem 1.1. In particular, verify the formulae

$$\bar{\mathfrak{h}}_r(\tau) = -\bar{\mathfrak{h}}(1-\tau), \qquad \tau \in [0,1],$$

$$\bar{\mathfrak{h}}_{\mathbf{P}_2 * \mathbf{P}_1}(\tau) = \begin{cases} 2\bar{\mathfrak{h}}_1(2\tau), & 0 \leqslant \tau < \frac{1}{2} \\ 2\bar{\mathfrak{h}}_2(2\tau - 1), & \frac{1}{2} \leqslant \tau < 1, \end{cases}$$

and use these formulae in the proof, where appropriate.

2. Discuss the following assertion: it is possible to add heat to a homogeneous fluid body in such a way that its temperature is lowered.

3. a. Verify for a homogeneous fluid body $\mathscr{F}$ that the heating $\bar{\mathfrak{h}}(\tau)$ at the instant τ for $(\bar{V}, \bar{\theta})$ can be written in the form

$$\bar{\mathfrak{h}}(\tau) = \tilde{\lambda}_p(\bar{V}(\tau), \bar{\theta}(\tau))\mathscr{p}^{\cdot}(\tau) + \mathscr{c}_p(\bar{V}(\tau), \bar{\theta}(\tau))\bar{\theta}^{\cdot}(\tau)$$

where $\tilde{\lambda}_p$ and $\mathscr{c}_p$ are given by

$$\tilde{\lambda}_p = \tilde{\lambda} \Big/ \frac{\partial \mathscr{p}}{\partial V},$$

$$\mathscr{c}_p = \mathscr{c} - \tilde{\lambda} \frac{\partial \mathscr{p} / \partial \theta}{\partial \mathscr{p} / \partial V},$$

and $\mathscr{p}^{\cdot}(\tau) = (d/d\tau)\mathscr{p}(\bar{V}(\tau), \bar{\theta}(\tau))$. [$\tilde{\lambda}_p$ is called the latent heat with respect to pressure and $\mathscr{c}_p$ the specific heat at constant pressure.]

   b. For an ideal gas, show that $\mathscr{c}_p(V,\theta) - \mathscr{c}(V,\theta) = \lambda$ for every state (V,θ), i.e., the difference in specific heats is the constant λ in the expression for $\tilde{\lambda}$.

   c. For an ideal gas with constant specific heats (i.e., with $\mathscr{c}(\theta) = c$ for all θ) verify that on each adiabat there holds $\mathscr{p}(V,\theta)V^{\gamma} = \text{constant}$, where $\gamma = \mathscr{c}_p / \mathscr{c}$ is the ratio of the specific heats.

4. Show that every internal energy function E and every entropy function S for an ideal gas have the respective forms

$$E(V,\theta)=J\int \mathfrak{d}(\theta)\,d\theta+\text{const.},$$

$$S(V,\theta)=\lambda\ln V+\int\frac{\mathfrak{d}(\theta)}{\theta}\,d\theta+\text{const.}$$

5. A homogeneous fluid body having pressure function

$$p(V,\theta)=\frac{R\theta}{V-b}-\frac{a}{V^2},$$

with R, b, and a positive constants, is called a Van der Waals fluid. What information do the First and Second Laws give about $\tilde{\lambda}$, $\mathfrak{a}$, E, and S for such a fluid?

6. How would you define the efficiency of a Carnot refrigerator? Analyze your definition using the First and Second Laws; in particular, obtain an analogue of Kelvin's formula for the efficiency of a Carnot heat engine, and discuss the question of adjusting the operating temperatures so as to increase the efficiency.

7. A Sargent cycle $\mathbb{S}$ for a homogeneous fluid body is described by giving four states (V_1,θ_1), (V_2,θ_2), (V_3,θ_3) and (V_4,θ_4), with θ_1 the largest and θ_3 the smallest of the four temperatures, and by joining them by paths $\mathbb{P}_{12}$, $\mathbb{P}_{23}$, $\mathbb{P}_{34}$, and $\mathbb{P}_{41}$ such that $\mathbb{P}_{12}$ and $\mathbb{P}_{34}$ are adiabats while $\mathbb{P}_{23}$ and $\mathbb{P}_{41}$ are isobars. Verify the following properties of a *Sargent cycle for an ideal gas* $\mathcal{G}$ with constant specific heat $\mathfrak{d}$, and sketch such a cycle:

a. $\dfrac{V_1}{V_4}=\dfrac{\theta_1}{\theta_4}$; $\dfrac{V_2}{V_3}=\dfrac{\theta_2}{\theta_3}$; $\dfrac{\theta_1}{\theta_4}=\dfrac{\theta_2}{\theta_3}$;

b. $\dfrac{W(\mathbb{S})}{JH^+(\mathbb{S})}=1-\dfrac{\theta_2}{\theta_1}$.

8. Suppose the pressure function for a homogeneous fluid body $\mathcal{F}$ satisfying the First and Second Laws has the form:

$$p(V,\theta)=f(V)+g(\theta),$$

with f differentiable and g twice differentiable. Verify the following:

a. $f'(V)<0$ for every $V>0$;

b. $\tilde{\lambda}(V,\theta)=\dfrac{\theta}{J}g'(\theta)$ for every (V,θ) in Σ;

c. $\mathfrak{a}(V,\theta)=\dfrac{\theta}{J}g''(\theta)V+\tilde{\mathfrak{d}}(\theta)$, where $\tilde{\mathfrak{d}}(\theta)>-\theta g''(\theta)V/J$ for every (V,θ) in Σ.

9. Let a homogeneous fluid body satisfy the First and Second Laws of thermodynamics.

a. Show that $\dfrac{\partial^2 p}{\partial\theta^2}(V,\theta)$ exists for every state (V,θ).

b. Verify the formula

$$\mathfrak{a}(V,\theta)=\frac{\theta}{J}\int\frac{\partial^2 p}{\partial\theta^2}(V,\theta)\,dV+\tilde{\mathfrak{d}}(\theta).$$

c. Justify or contradict the following assertion: the laws of thermodynamics along with the pressure function p of a homogeneous fluid body determine $\tilde{\Lambda}$ completely and determine s, E, and S each to within a function of temperature alone.

Chapter II

1. Let $(\Sigma_{\mathscr{F}}, \Pi_{\mathscr{F}})$ be the system with perfect accessibility associated with a homogeneous fluid body $\mathscr{F}$. Verify that the formula

$$d(\pi_t, \sigma) = t, \ (\pi_t, \sigma) \in \Pi_{\mathscr{F}} \diamond \Sigma_{\mathscr{F}}$$

defines an action for $(\Sigma_{\mathscr{F}}, \Pi_{\mathscr{F}})$. Give examples of two processes (π_{t_1}, σ) and (π_{t_2}, σ) such that

$$\mathscr{D}(\pi_{t_1}) = \mathscr{D}(\pi_{t_2}),$$

$$\rho_{\pi_{t_1}} = \rho_{\pi_{t_2}},$$

but

$$d(\pi_{t_1}, \sigma) \neq d(\pi_{t_2}, \sigma).$$

2. a. Let π_1, π_2, and π_3 be process generators of a system with perfect accessibility (Σ, Π) such that $(\pi_3\pi_2)\pi_1$ and $\pi_3(\pi_2\pi_1)$ are both defined. Show that

$$\mathscr{D}((\pi_3\pi_2)\pi_1) = \mathscr{D}(\pi_3(\pi_2\pi_1)),$$

$$\rho_{(\pi_3\pi_2)\pi_1} = \rho_{\pi_3(\pi_2\pi_1)},$$

and, for every action a for (Σ, Π),

$$a((\pi_3\pi_2)\pi_1, \sigma) = a(\pi_3(\pi_2\pi_1), \sigma)$$

for each $\sigma \in \mathscr{D}((\pi_3\pi_2)\pi_1)$. Are we then justified in asserting that $(\pi_3\pi_2)\pi_1$ equals $\pi_3(\pi_2\pi_1)$?

b. If both $\pi_3\pi_2$ and $\pi_2\pi_1$ are defined, does it follow that $\pi_3(\pi_2\pi_1)$ and $(\pi_3\pi_2)\pi_1$ are also defined?

Chapter III

1. Show that for each process generator (π_1, π_2) of $\mathscr{S}_1 \times \mathscr{S}_2$, the ranges of (π_1, π_2), π_1, and π_2 are related by

$$\mathscr{R}(\pi_1, \pi_2) = \mathscr{R}(\pi_1) \times \mathscr{R}(\pi_2).$$

2. On the basis of the results in Chapter III, discuss the following assertion: work and heat play symmetrical roles in the First Law.

Chapter IV

1. For the Carnot refrigerator in Figure 21, verify the relation $V_1V_3 = V_2V_4$ as well as the formula (3.4).

2. For the Sargent cycle $\mathbb{S}$ described in problem 7, Chapter I, give a description of the accumulation function $\theta \mapsto H_{\mathscr{G}}(\mathbb{S}, \theta)$. Be as specific as possible by sketching the graph of this function and giving quantitative information about the graph.

3. Suppose a cycle (π, σ) of $\mathscr{S}$ has accumulation function of the form

$$H_{\mathscr{S}}\left(\pi, \sigma, \varphi_{\mathscr{G}}^{-1}(\theta)\right) = \begin{cases} 0, & 0 < \theta < \theta_1 \\ 1, & \theta_1 \leqslant \theta < \theta_2 \\ A, & \theta_2 \leqslant \theta < \theta_3 \\ 1, & \theta_3 \leqslant \theta. \end{cases}$$

If $\mathscr{S} \times \mathscr{G}$ obeys the Second Law, show that A must be negative and verify that

$$A \leqslant \frac{\dfrac{1}{\theta_2} - \left(\dfrac{1}{\theta_1} + \dfrac{1}{\theta_3}\right)}{\dfrac{1}{\theta_2} - \dfrac{1}{\theta_3}} \leqslant -\frac{\theta_3(\theta_2 - \theta_1)}{\theta_1(\theta_3 - \theta_2)}.$$

4. Prove Lemma 6.3 when the empirical temperature scale has range equal to a proper subset of $\mathbb{R}^{++}$. [You will have to extend $\theta \mapsto H_{\mathscr{G}}(\pi, \sigma, \varphi^{-1}(\theta))$ from rng φ to all of $\mathbb{R}^{++}$ in order to use the arguments in the text.]

5. Supply statements in terms of accumulation functions of versions 1 and 3 of the Second Law given in the introduction to Chapter IV.

6. If both $\mathscr{S}$ and $\mathscr{S} \times \mathscr{G}$ obey the Second Law, show that $\mathscr{S}$ obeys (4.3).

7. Prove that equality holds in (7.3) if and only if both (7.4) and (7.5) are satisfied.

Chapter V

1. Let a be an action with the Clausius property at a state σ° of a system with perfect accessibility (Σ, Π). Suppose that for each process (π, σ°) there is a process $(\tilde{\pi}, \rho_\pi\sigma^\circ)$ such that

$$\rho_{\tilde{\pi}}(\rho_\pi\sigma^\circ) = \sigma^\circ$$

and

$$a(\pi, \sigma^\circ) + a(\tilde{\pi}, \rho_\pi\sigma^\circ) = 0.$$

Show that $s(\sigma^\circ, \sigma^\circ) = 0$ and describe the collection of upper potentials for a which vanish at σ°.

2. Find two examples of actions for a homogeneous fluid body which have the dissipation property but not the conservation property at a state. In each example, find a lower potential for the action.

3. A *semi-system* is a pair (Σ, Π) which satisfies all the conditions in the definition of a system with perfect accessibility except that the condition

   "(S1): *for each* σ *in* Σ, the set
   $$\Pi\sigma^\circ := \{ \rho_\pi \sigma \mid \pi \in \Pi, \sigma \in \mathscr{D}(\pi) \}$$
   equals Σ,"

   is replaced by

   "(S1)′: *there exists* σ° *in* Σ such that
   $$\Pi\sigma^\circ := \{ \rho_\pi \sigma^\circ \mid \pi \in \Pi, \sigma^\circ \in \mathscr{D}(\pi) \}$$
   equals Σ."

   An action for a semi-system is defined exactly as for a system, as are upper potentials for actions. Show that an action a for a semi-system has an upper potential if and only if a satisfies: for each σ in Σ, the set $\{ a(\pi, \sigma^\circ) \mid \rho_\pi \sigma^\circ = \sigma \}$ is bounded above.

4. Consider systems with perfect accessibility (Σ_1, Π_1), (Σ_2, Π_2), and (Σ, Π) for which each σ in Σ determines a unique state σ_1 in Σ_1 and a unique state σ_2 in Σ_2, and every state in Σ_1 and every state in Σ_2 is determined by a state in Σ. Assume that Π_1, Π_2, and Π are related in a similar way. (This situation might arise if (Σ, Π) describes a body in space and $(\Sigma_1, \Pi_1), (\Sigma_2, \Pi_2)$ each describes a subbody of the given body.) The relation between these systems can be described by functions ω_1, ω_2, μ_1, and μ_2, as shown below, each a surjection:

   $$\Sigma_1 \xleftarrow{\omega_1} \Sigma \xrightarrow{\omega_2} \Sigma_2 \qquad \Pi_1 \xleftarrow{\mu_1} \Pi \xrightarrow{\mu_2} \Pi_2$$

   Suppose further that for each π, π', π'' in Π, with $\pi''\pi'$ defined, for each σ in $\mathscr{D}(\pi)$, and for each $i = 1,2$ there hold
   $$\omega_i(\mathscr{D}(\pi)) = \mathscr{D}(\mu_i(\pi)),$$
   $$\rho_{\mu_i(\pi)} \omega_i(\sigma) = \omega_i(\rho_\pi \sigma),$$
   $$\mu_i(\pi''\pi') = \mu_i(\pi'')\mu_i(\pi').$$

   a. If a_1 is an action for (Σ_1, Π_1), a_2 is an action for (Σ_2, Π_2), and $(\pi, \sigma) \mapsto a(\pi, \sigma)$ is the function defined by
   $$a(\pi, \sigma) = a_1(\mu_1 \pi, \omega_1 \sigma) + a_2(\mu_2 \pi, \omega_2 \sigma) \qquad (*)$$
   for each $(\pi, \sigma) \in \Pi \diamond \Sigma$, show that each term in the right hand side of $(*)$ is meaningful and that a is an action for (Σ, Π).

   b. If σ° in Σ is such that a_1 and a_2 have the Clausius property at $\omega_1 \sigma^\circ$ and $\omega_2 \sigma^\circ$, respectively, then show that a has the Clausius property at σ°.

   c. If A_1 and A_2 are upper potentials for a_1 and a_2, respectively, then show that $\sigma \mapsto A_1(\omega_1(\sigma)) + A_2(\omega_2(\sigma))$ is an upper potential for a.

   d. Let a_1, a_2, and σ° be as in part b, let A°, A_1°, and A_2° denote the smallest upper potentials for a, a_1, and a_2, respectively, which vanish at σ°, $\omega_1(\sigma^\circ)$, and $\omega_2(\sigma^\circ)$, respectively. Show that
   $$A^\circ(\sigma) \leqslant A_1^\circ(\omega_1(\sigma)) + A_2^\circ(\omega_2(\sigma))$$
   for every σ in Σ.

Chapter VI

1. Reformulate the entire discussion on elastic–perfectly plastic filaments using the variables ℓ_R and ℓ_e given by

$$\ell_R = \ell_R(\ell, f) = \ell - \frac{f}{\beta}$$

$$\ell_e = \ell_e(\ell, f) = \frac{f}{\beta}.$$

In particular, describe the state space Σ in the $\ell_R - \ell_e$ plane, and describe the set $\mathscr{L}^\circ$ of Helmholtz free energy functions using the variables ℓ_R and ℓ_e in place of ℓ and f.

2. Use relations (4.6) and (4.7) to show that the paths $\Gamma(\pi_t, (\ell, f))$ and $\Gamma(\pi_{\tilde{t}}, (\ell, f)$ are the same if

$$\pi_{\tilde{t}}(\tau) = \pi_t(\psi(\tau)), \qquad 0 \leqslant \tau < \tilde{t} = \psi^{-1}(t),$$

where ψ is a strictly increasing smooth function from $[0, \tilde{t}]$ onto $[0, t]$.

Chapter VII

1. Give detailed proofs of Theorems 3.1 and 3.2.

2. Verify the italicized statement on efficiency of processes given in the last paragraph of Section 3.

Index

Undergraduate Texts in Mathematics

continued from ii

Owan: A First Course in the Mathematical Foundations of Thermodynamics
1984. xvii, 178 pages. 52 illus.

Prenowitz/Jantosciak: Join Geometrics: A Theory of Convex Set and Linear Geometry.
1979. xxii, 534 pages. 404 illus.

Priestly: Calculus: An Historical Approach.
1979, xvii, 448 pages. 335 illus.

Protter/Morrey: A First Course in Real Analysis.
1977. xii, 507 pages. 135 illus.

Ross: Elementary Analysis: The Theory of Calculus.
1980. viii, 264 pages. 34 illus.

Sigler: Algebra.
1976. xii, 419 pages. 27 illus.

Simmonds: A Brief on Tensor Analysis.
1982. xi, 92 pages. 28 illus.

Singer/Thorpe: Lecture Notes on Elementary Topology and Geometry.
1976. viii, 232 pages. 109 illus.

Smith: Linear Algebra.
1978. vii, 280 pages. 21 illus.

Smith: Primer of Modern Analysis
1983. xiii, 442 pages. 45 illus.

Thorpe: Elementary Topics in Differential Geometry.
1979. xvii, 253 pages. 126 illus.

Troutman: Variational Calculus with Elementary Convexity.
1983. xiv, 364 pages. 73 illus.

Whyburn/Duda: Dynamic Topology.
1979. xiv, 338 pages. 20 illus.

Wilson: Much Ado About Calculus: A Modern Treatment with Applications Prepared for Use with the Computer.
1979. xvii, 788 pages. 145 illus.

Printed in the USA
CPSIA information can be obtained
at www.ICGtesting.com
LVHW021111300324
775943LV00001B/33